Applied
General
Mathematics

DELMAR PUBLISHERS INC.
2 Computer Drive-West Box 15-015
Albany, New York 12212

Applied General Mathematics

Robert Smith

COPYRIGHT © 1982
BY DELMAR PUBLISHERS INC.

All rights reserved. No part of this work covered by the copyright hereon may be reproduced or used in any form or by any means — graphic, electronic, or mechanical, including photocopying, recording, taping, or information storage and retrieval systems — without written permission of the publisher.

10 9 8 7 6 5 4

LIBRARY OF CONGRESS CATALOG CARD NUMBER: 79-51586
ISBN: 0-8273-1674-7

Printed in the United States of America
Published simultaneously in Canada
by Nelson Canada,
A Division of International Thomson Limited

Preface

Mathematical skills are encountered in all career fields. *Applied General Mathematics* is intended for individuals who need to develop a practical understanding of fundamental mathematical skills. The mathematics presented is geared to students' interest levels and future needs. The content can be used with a wide range of students at various age levels and stages in career development.

The author has made an intensive study of resource materials from various career fields. He has included many practical problems relevant to the mathematical concepts presented. Practical applications require the student to become involved in the type of thinking significant in real situations. Problems often contain an explanation of the career as well as direct student involvement in the subject matter.

Presentation of basic concepts is accompanied by examples and practices with answers. Mathematical principles and techniques are reinforced by immediate drill exercises. This ensures that the student is knowledgeable about the concept before using it in application situations. Unit reviews and practical applications are found at the end of each unit. Many problems require the student to work with illustrations and charts such as are found in handbooks and engineering drawings. Charts of equivalent units of measure are found in the appendix.

An Instructor's Guide (including answers to all exercises and complete solutions for the practical applications) and a Test Booklet complement the text.

Robert D. Smith has experience in the manufacturing industry, in education, and in publishing. He held positions as tool designer, quality control engineer, and chief manufacturing engineer prior to teaching. Mr. Smith taught applied mathematics, physics, and industrial materials and processes on the secondary school level at the Al Prince Vocational-Technical School in Hartford, Connecticut. He has been an author and consulting editor on other texts. Mr. Smith is presently a faculty member of the Vocational-Technical Education Department at Central Connecticut State College, New Britain, Connecticut. He has been involved in several professional organizations in his field of interest, including the American Technical Education Association and the Society of Manufacturing Engineers.

A series of Practical Problems in Mathematics publications is available which provides additional problem material to meet the needs in specific occupational career areas. The workbooks in this series are:

PRACTICAL PROBLEMS IN MATHEMATICS FOR
    AUTOMOTIVE TECHNICIANS
    CARPENTERS
    CONSUMERS
    COSMETOLOGY
    ELECTRICIANS
    GRAPHIC ART
    HEATING AND COOLING TECHNICIANS
    MACHINISTS
    MASONS
    MECHANICAL DRAFTING
    OFFICE WORKERS
    PRINTERS
    SHEET METAL TECHNICIANS
    THE METRIC SYSTEM
    WELDERS

Other comprehensive texts available from Delmar Publishers Inc.
    BASIC MATHEMATICS SIMPLIFIED
    BUSINESS MATHEMATICS
    MATHEMATICS FOR AUTO MECHANICS
    MATHEMATICS FOR CAREERS
    MATHEMATICS FOR CARPENTERS
    MATHEMATICS FOR HEALTH CAREERS
    MATHEMATICS FOR MACHINE TECHNOLOGY
    MATHEMATICS FOR PLUMBERS AND PIPEFITTERS
    MATHEMATICS FOR SHEET METAL FABRICATION
    MATHEMATICS FOR THE SHOP
    MERCHANDISING MATHEMATICS

Contents

SECTION 1 Whole Numbers
- Unit 1   Introduction to Whole Numbers 1
- Unit 2   Addition of Whole Numbers 8
- Unit 3   Subtraction of Whole Numbers 14
- Unit 4   Multiplication of Whole Numbers 21
- Unit 5   Division of Whole Numbers 33
- Unit 6   Order of Operations 43

SECTION 2 Common Fractions
- Unit 7   Introduction to Common Fractions 51
- Unit 8   Addition of Common Fractions 66
- Unit 9   Subtraction of Common Fractions 79
- Unit 10 Multiplication of Common Fractions 90
- Unit 11 Division of Common Fractions 103
- Unit 12 Combined Operations with Common Fractions 116

SECTION 3 Decimal Fractions
- Unit 13 Introduction to Decimal Fractions 125
- Unit 14 Equivalent Decimal and Common Fractions 132
- Unit 15 Addition and Subtraction of Decimal Fractions 141
- Unit 16 Multiplication of Decimal Fractions 152
- Unit 17 Division of Decimal Fractions 165
- Unit 18 Powers and Roots of Decimal Fractions 176
- Unit 19 Combined Operations with Decimal Fractions 193

SECTION 4 Percentage, Statistical Measures, and Graphs
- Unit 20 Percents and Simple Percentage Problems 207
- Unit 21 More Complex Percentage Problems 222
- Unit 22 Statistical Measures 228
- Unit 23 Bar Graphs 240
- Unit 24 Line Graphs 253

SECTION 5 Measure
- Unit 25 English Units of Linear Measure 274
- Unit 26 Metric Units of Linear Measure 291
- Unit 27 Equivalent Units of Area Measure 306
- Unit 28 Equivalent Units of Volume Measure 319
- Unit 29 Equivalent Units of Capacity and Mass Measure 330

Appendix 341

Index 347

SECTION 1
Whole Numbers

UNIT 1
Introduction to Whole Numbers

OBJECTIVES

After studying this unit you should be able to

- Express the digit place values of whole numbers.
- Write whole numbers in expanded form.
- Read and write whole numbers.

All occupations, from the least to the most highly skilled, require the use of mathematics. The basic operations of mathematics are addition, subtraction, multiplication, and division. These operations are based on the decimal system. Therefore, it is important that you understand the structure of the decimal system before doing the basic operations.

The development of the decimal system can be traced back many centuries. In ancient times small numbers were counted by comparing the number of objects with the number of fingers. To count larger numbers pebbles might be used. One pebble represented one counted object. Counting could be done more quickly when the pebbles were placed in groups, generally ten pebbles in each group. Our present number system, the decimal system, is based on this ancient practice of grouping by ten.

PLACE VALUE

In the decimal system, ten number symbols or digits are used. The digits, 0, 1, 2, 3, 4, 5, 6, 7, 8, and 9, can be arranged to represent any number. The value expressed by each digit depends on its position in the written number. This value is called the *place value*. The chart shows the place value for each digit in the number 2 452 678 932.

BILLIONS	HUNDRED MILLIONS	TEN MILLIONS	MILLIONS	HUNDRED THOUSANDS	TEN THOUSANDS	THOUSANDS	HUNDREDS	TENS	UNITS
2	4	5	2	6	7	8	9	3	2

The digit on the far right is in the units place. The digit second from the right is in the tens place. The digit third from the right is in the hundreds place. The value of each place is ten times the value of the place directly to its right.

Example: Write the place value of the 3 in the number 63 125.

Start with the digit on the far right.
The 5 is in the units place.
The 2 is in the tens place.
The 1 is in the hundreds place.
The 3 is in the thousands place.
The 6 is in the ten thousands place.

TEN THOUSANDS	THOUSANDS		HUNDREDS	TENS	UNITS
6	3		1	2	5

Thousands *Ans*

Practice

Write the place value of the underlined digit in each number.

1. 23 <u>1</u>64 . . . . Hundreds
2. 52<u>3</u> . . . . Units
3. <u>1</u>43 892 . . . . Hundred Thousands
4. 8<u>9</u> 874 726 . . . . Millions
5. 7 6<u>2</u>3 . . . . Tens

EXPANDING WHOLE NUMBERS

The number 64 is a simplified and convenient way of writing 6 tens plus 4 ones. In its expanded form, 64 is shown as $(6 \times 10) + (4 \times 1)$.

> Example: Write the number 382 in expanded form.
>
> 382 = 3 hundreds plus 8 tens plus 2 ones
> 382 = $(3 \times 100) + (8 \times 10) + (2 \times 1)$ *Ans*

Practice

Write each number in expanded form.

1. 7 028 . . . . $(7 \times 1\,000) + (0 \times 100) + (2 \times 10) + (8 \times 1)$
2. 12 . . . . $(1 \times 10) + (2 \times 1)$
3. 234 . . . . $(2 \times 100) + (3 \times 10) + (4 \times 1)$
4. 86 279 . . . . $(8 \times 10\,000) + (6 \times 1\,000) + (2 \times 100) + (7 \times 10) + (9 \times 1)$
5. 345 . . . . $(3 \times 100) + (4 \times 10) + (5 \times 1)$

Exercise 1-1

Write the place value for the specified digit of each number given in the tables.

	DIGIT	NUMBER	PLACE VALUE	
1.	7	6 732	Hundreds	*Ans*
2.	3	139		
3.	6	16 137		
4.	1	4 531		
5.	3	136 805		
6.	2	427		
7.	9	9 732 500		
8.	5	4 578 190		

	DIGIT	NUMBER	PLACE VALUE
9.	1	10 070	
10.	0	20 123	
11.	9	98	
12.	7	782 944	
13.	5	153 400	
14.	9	98 600 057	
15.	2	378 072	
16.	4	43 728	

4 Section 1 Whole Numbers

Write each whole number in expanded form.

17. 857 = (8 × 100) + (5 × 10) + (7 × 1) *Ans*
18. 32
19. 942
20. 1 372
21. 10
22. 5 047
23. 379
24. 23 813
25. 504
26. 6 376
27. 333
28. 59
29. 600
30. 685 412
31. 90 507
32. 7 500 000
33. 178 512
34. 70 001
35. 234 123
36. 17 643 000
37. 428 000 975

Write each expanded whole number in its simplified form.

38. (5 × 10) + (3 × 1)   53 *Ans*
39. (7 × 10) + (4 × 1)
40. (3 × 100) + (6 × 10) + (2 × 1)
41. (8 × 100) + (3 × 10) + (0 × 1)
42. (6 × 100) + (0 × 10) + (8 × 1)
43. (4 × 100) + (6 × 10) + (0 × 1)
44. (2 × 1 000) + (3 × 100) + (9 × 10) + (4 × 1)
45. (7 × 1 000) + (5 × 100) + (0 × 10) + (1 × 1)
46. (1 × 1 000) + (0 × 100) + (0 × 10) + (0 × 1)
47. (9 × 10 000) + (8 × 1 000) + (1 × 100) + (3 × 10) + (2 × 1)
48. (4 × 10 000) + (0 × 1 000) + (0 × 100) + (9 × 10) + (9 × 1)
49. (6 × 10 000) + (7 × 1 000) + (6 × 100) + (5 × 10) + (4 × 1)
50. (8 × 1 000) + (3 × 100) + (9 × 10) + (0 × 1)

READING AND WRITING WHOLE NUMBERS

The numbers 0, 1, 2, 3, 4, 5, 6, 7, 8, 9, 10, 11, 12, etc. are *whole numbers*. Beginning with 1, these are the numbers that are used for counting and are called *counting numbers*. There is no greatest counting number since 1 can be added to any number to obtain a larger number. Whole numbers indicate whole or complete quantities such as $15 which represents 15 whole dollars or 50 bolts which represents 50 whole bolts.

To read a whole number or to write a whole number as a word statement the place value of each digit of the number must be considered. Starting from the far right (units place) and going to the left, group each set of three digits. Leave a space between each set. The number is read or written as a word statement from left to right. Each set of three digits uses the proper name according to the grouping. When writing the number as a word statement a hyphen is used between the first and second words of the numbers from twenty-one through ninety-nine.

Example: Write 8627139364 as a word statement.

GROUP NAME	BILLION			MILLION			THOUSAND					
PLACE VALUE	HUNDRED BILLIONS	TEN BILLIONS	BILLIONS	HUNDRED MILLIONS	TEN MILLIONS	MILLIONS	HUNDRED THOUSANDS	TEN THOUSANDS	THOUSANDS	HUNDREDS	TENS	UNITS
NUMBER			8	6	2	7	1	3	9	3	6	4

Start at the right and group each set of three digits.
Write the number as a word statement from left to right.
Each set of three digits uses the proper name according to the grouping.
Eight billion, six hundred twenty-seven million, one hundred thirty-nine thousand, three hundred sixty-four *Ans*

Practice

Write each number as a word statement.

1. 9 364. . . . . Nine thousand, three hundred sixty-four
2. 139 364. . . . . One hundred thirty-nine thousand, three hundred sixty-four
3. 7 139 364 . . . . .Seven million, one hundred thirty-nine thousand, three hundred sixty-four

Exercise 1-2

Write each whole number as a word statement.

1. 829 = Eight hundred twenty-nine *Ans*
2. 942
3. 53
4. 14
5. 301
6. 5 342
7. 9 404
8. 13 426
9. 123 130
10. 672 014
11. 2 850 673
12. 6 900 049
13. 23 000 000
14. 870 314 012
15. 333 333 333
16. 4 173 932 512
17. 5 072 014 017

Write each word statement as a whole number.

18. Seven thousand, three hundred thirty-one = 7 331 *Ans*
19. Three hundred forty-three
20. Seven hundred twenty-five

6 Section 1 Whole Numbers

21. Five thousand, seven hundred eighty-two
22. Thirteen thousand, nine hundred sixty-six
23. Seventy thousand, five hundred seventy-three
24. Nine hundred fifty-two thousand, seven hundred
25. One hundred thousand, five
26. Three million, eight hundred thousand, three hundred thirty-two
27. Eighteen million, four hundred eighty-seven thousand, three hundred seventy-six

UNIT REVIEW

Exercise 1-3

Write the place value for the specified digit of each number given in the table.

	DIGIT	NUMBER	PLACE VALUE
1.	3	6 938	
2.	5	519	
3.	7	27 043	
4.	6	168 700	
5.	2	9 842	
6.	2	2 815 000	

	DIGIT	NUMBER	PLACE VALUE
7.	8	5 810 612	
8.	0	60 443	
9.	7	1 075	
10.	1	513 600	
11.	9	98 033	
12.	4	46 718 000	

Write each whole number in expanded form.

13. 48
14. 319
15. 2 146
16. 776
17. 13 692
18. 863
19. 22
20. 856 241
21. 40 007
22. 6 600 000
23. 38 694 000
24. 655 200 045

Write each expanded whole number in its simplified form.

25. (7 × 10) + (4 × 1)
26. (2 × 100) + (8 × 10) + (5 × 1)
27. (5 × 10 000) + (0 × 1 000) + (8 × 100) + (1 × 10) + (0 × 1)
28. (9 × 1 000) + (9 × 100) + (0 × 10) + (3 × 1)
29. (1 × 10 000) + (1 × 1 000) + (2 × 100) + (0 × 10) + (9 × 1)
30. (8 × 10 000) + (0 × 1 000) + (0 × 100) + (7 × 10) + (8 × 1)

Write each whole number as a word statement.

31. 767
32. 38
33. 19
34. 4 164
35. 17 608
36. 132 260
37. 3 812 500
38. 54 700 000
39. 70 080 505
40. 613 306 000
41. 104 040 040
42. 9 720 610 200

Write each word statement as a whole number.

43. Nine thousand, four hundred twenty-two
44. Seven hundred thirty-eight
45. Eighty thousand, five hundred twelve
46. Four hundred sixty-four thousand, one hundred thirty-three
47. Two million, seven hundred thousand, four hundred forty-five
48. Sixty-six million, one hundred thousand, four hundred thirty
49. Thirty-four thousand, three hundred four
50. One hundred fifty-nine million, two hundred forty-two

UNIT 2
Addition of Whole Numbers

OBJECTIVES

After studying this unit you should be able to

- Arrange and add whole numbers.
- Solve practical problems by addition.

A contractor determines the cost of materials in a building. A salesperson charges a customer for the total cost of a number of purchases. An air conditioning and refrigeration technician finds lengths of duct needed. These people are using addition. Practically every occupation requires daily use of addition.

DEFINITIONS AND PROPERTIES OF ADDITION

The quantities to be added are called *addends*. The result of the addition (answer) is called the *sum*. The *plus sign* (+) indicates addition.

Numbers can be added in any order. The same sum is obtained regardless of the order in which the numbers are added. For example, 2 + 4 + 3 may be added in either of the following ways:

$$2 + 4 + 3 = 9 \quad \text{or} \quad 3 + 4 + 2 = 9$$

The numbers can also be grouped in any way and the sum is the same.

$$\begin{array}{cc} (2 + 4) + 3 & 2 + (4 + 3) \\ 6 + 3 = 9 & \text{or} \quad 2 + 7 = 9 \end{array}$$

PROCEDURE FOR ADDING WHOLE NUMBERS

Write the numbers to be added under each other. Place the units digits under the units digit, the tens digits under the tens digit, etc. Add each column of numbers starting from the column on the right (units column). If the sum of any column is ten or more, write the last digit of the sum in the answer. Mentally add the rest of the number to the next column. Continue the same procedure until all columns are added.

The sum can be checked by adding the numbers in the opposite direction. If the numbers were added from top to bottom, check by adding from bottom to top.

Example: Add and check.

$$875 + 2\,318 + 27 + 5\,079 + 8$$

Write the numbers under each other, placing digits in proper place positions.

Add the numbers in the units column from top to bottom.
$5 + 8 + 7 + 9 + 8 = 37$
Write 7 in the answer.

```
    8 7 5
  2 3 1 8
      2 7
  5 0 7 9
+       8
  8 3 0 7  Ans
```

Add the 3 to the numbers in the tens column.
$(3) + 7 + 1 + 2 + 7 = 20$
Write 0 in the answer.

Add the 2 to the numbers in the hundreds column.
$(2) + 8 + 3 + 0 = 13$
Write 3 in the answer.

Add the 1 to the numbers in the thousands column.
$(1) + 2 + 5 = 8$
Write 8 in the answer.

Check the answer. The numbers in each column were added from top to bottom. Check by adding the numbers in each column from bottom to top.

```
  8 3 0 7  Ck
    8 7 5
  2 3 1 8
      2 7
  5 0 7 9
+       8
  8 3 0 7  Ans
```

Practice

Add and check.

1. 82 + 17 99
2. 49 + 7 56
3. 304 + 43 + 7 354
4. 24 + 592 616
5. 60 + 282 + 17 359

Exercise 2-1

Add and check.

1. 43
   + 26

2. 37
   + 81

3. 19
   + 23

4. 75
   + 89

5. 96
   + 25

6. 312
   + 17

7. 953
   + 38

8. 347
   + 85

9. 634
   + 853

10. 4 767
    + 9 328

11. 317
    8
    + 19

12. 633
    + 349

13. 131
    9
    + 4 732

14. 6 737
    3 519
    + 8 180

15. 9 734
    10 505
    + 91 613

16. 7 + 18
17. 33 + 88
18. 53 + 89 + 718
19. 531 + 12 + 9 515
20. 7 + 13 700 + 88
21. 66 112 + 313 + 9 556
22. 17 392 + 2 085 + 1 670 + 13 + 83 006
23. 38 + 55 404 + 132 997 + 60 004 + 8
24. 18 768 + 3 023 + 7 787 030 + 38 + 544

PRACTICAL APPLICATIONS

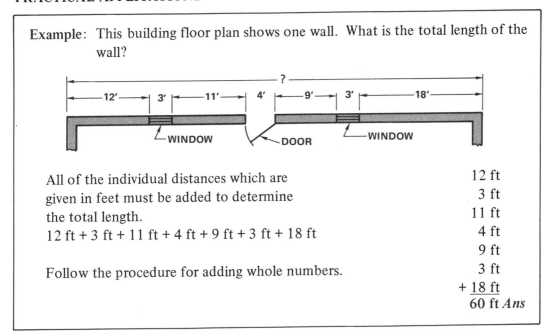

Example: This building floor plan shows one wall. What is the total length of the wall?

All of the individual distances which are given in feet must be added to determine the total length.

12 ft + 3 ft + 11 ft + 4 ft + 9 ft + 3 ft + 18 ft

Follow the procedure for adding whole numbers.

```
  12 ft
   3 ft
  11 ft
   4 ft
   9 ft
   3 ft
+ 18 ft
  60 ft Ans
```

Exercise 2-2

1. A contractor estimates the costs of remodeling a home as follows: masonry $525; lumber $710; hardware $58; trim $47; paint $108; and labor $2 845. What is the total estimated cost?

2. A dietitian plans and computes the number of calories in a meal. The meal contains: chicken soup, 44 calories; roast beef, 525 calories; potato, 112 calories; peas, 47 calories; bread, 60 calories; milk, 158 calories; cookies, 134 calories. How many total calories are contained in this meal?

3. A machinist checks distances on a steel base plate. Use the diagram of the plate and find, in centimetres, distances A, B, C, and D.

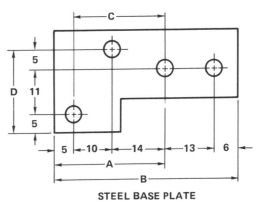

STEEL BASE PLATE

12 Section 1 Whole Numbers

4. A common unit of electrical power is the watt. A certain electrical circuit has one 200-watt lamp, two 100-watt lamps, three 75-watt lamps, and one 60-watt lamp. Determine the total number of watts in the circuit.

5. In a printing shop a press operator prints 9 552 letterheads on Monday, 11 768 on Tuesday, 10 855 on Wednesday, 8 970 on Thursday, and 12 314 on Friday. How many letterheads are printed during the week?

6. A landscaper is contracted to prepare and seed a plot of land. The cost of the job is determined by the total number of square feet in the plot. What is the total number of square feet in the plot shown?

LOT 5 19 250 sq ft		LOT 4 18 130 sq ft
LOT 1 17 870 sq ft	LOT 2 15 265 sq ft	LOT 3 16 035 sq ft

PLOT PLAN

7. In preparing a cake mix, a baker uses 12 pounds of flour, 6 pounds of shortening, 15 pounds of sugar, 10 pounds of milk, 7 pounds of whole eggs, and a total of 2 pounds of flavoring, salt and baking powder. What is the total weight of the mix?

8. The weight of each piece of steel required by a welder for a certain job is shown on the chart.

PLATE STEEL (Pounds)	CHANNEL IRON (Pounds)	ANGLE IRON (Pounds)	PIPE (Pounds)	I BEAM (Pounds)
1 775	305	88	5	352
432	57	276	32	47
67	8	104	97	1 022
2 985	108			213
	96			

a. Find the number of pounds of plate steel used.
b. Find the number of pounds of channel iron used.
c. Find the number of pounds of angle iron used.
d. Find the number of pounds of pipe used.
e. Find the number of pounds of I beam used.
f. Find the number of pounds of steel used for the job.

9. An appliance retailer's sales for the first six months of last year are Jan, $37 540; Feb, $32 178; Mar, $40 357; Apr, $42 116; May, $48 050; Jun, $53 730. What are the total sales for the six-month period?

10. A building foundation is shown. Find the total distance, in metres, around the outside edge (perimeter) of the building.

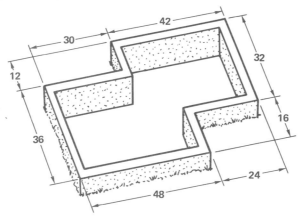

11. During a three-month period, a drafter works on five design projects. The drafter spends 96 hours on Project A, 43 hours on Project B, 178 hours on Project C, 120 hours on Project D, and 85 hours on Project E. Find the total hours worked in three months.

12. An estimator for a construction firm finds the cost of laying sidewalks for a manufacturing plant. The sidewalks are shown in the plan. What is the total length of the sidewalks?

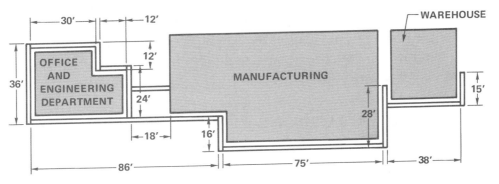

UNIT 3
Subtraction of Whole Numbers

OBJECTIVES

After studying this unit you should be able to
- Arrange and subtract whole numbers.
- Solve practical problems by subtraction.
- Solve problems using addition and subtraction.

A plumber uses subtraction to compute material requirements of a job. A machinist determines locations of holes to be drilled. A retail clerk inventories merchandise. An electrician estimates the profit of a wiring installation. Subtraction has many on-the-job applications.

DEFINITIONS

Subtraction is the operation which determines the difference between two quantities. It is the *inverse* or opposite of addition. The quantity subtracted is called the *subtrahend*. The quantity from which the subtrahend is subtracted is called the *minuend*. The result of the subtraction operation is called the *difference* or *remainder*. The *minus sign* (–) indicates subtraction.

PROCEDURE FOR SUBTRACTING WHOLE NUMBERS

Write the number to be subtracted (subtrahend) under the number from which it is subtracted (minuend). Place the units digit under the units digit, the tens digit under the tens digit, etc. Subtract each column of numbers starting from the right (units column).

If the digit in the subtrahend represents a value greater than the value of the corresponding digit in the minuend, it is necessary to regroup. Regroup the number in the minuend by taking 1 from the number in the next higher place and adding 10 to the number in the place directly to the right. The value of the minuend remains unchanged. The value represented by each digit of a number is 10 times the value represented by the digit directly to its right. For example, 85 is a convenient way of writing 8 tens plus 5 ones, $(8 \times 10) + (5 \times 1)$. The 5 in 85 can be increased to 15 by taking 1 from the 8 without changing the value of 85. This process is called *regrouping*, since 8 tens and 5 units $(80 + 5)$ is regrouped as 7 tens and 15 units $(70 + 15)$.

The difference can be checked by adding the difference to the subtrahend. The sum should equal the minuend. If the sum does not equal the minuend, go over the operation to find the error.

Example: Subtract and check.

$9\ 756 - 582$

Write the subtrahend 582 under the minuend 9 756. Place the digits in the proper place positions.

$$\begin{array}{r} 9\ 756 \\ -\ \ 582 \\ \hline 9\ 174\ Ans \end{array}$$

Subtract the units.
$6 - 2 = 4$
Write 4 in the answer.

Subtract the tens.
Since 8 is larger than 5, regroup the number.
$(9 \times 1\ 000) + (6 \times 100) + (15 \times 10) + (6 \times 1)$
Subtract the tens.
$15 - 8 = 7$
Write 7 in the answer.

Subtract the hundreds.
$6 - 5 = 1$
Write 1 in the answer.

Subtract the thousands.
$9 - 0 = 9$
Write 9 in the answer.

Check the answer. Adding the answer 9 174 and the subtrahend 582 equals the minuend 9 756.

$$\begin{array}{r} 9\ 174 \\ +\ \ 582 \\ \hline 9\ 756\ Ck \end{array}$$

Practice

Subtract and check.
1. 389 − 43 346
2. 602 − 21 581
3. 3 491 − 43 .... 3 448
4. 5 492 − 463 .... 5 029
5. 600 − 79 521

Exercise 3-1

Subtract and check.

1. 33
   − 21

2. 97
   − 65

3. 89
   − 5

4. 37
   − 10

5. 67
   − 23

6. 28
   − 19

7. 46
   − 38

8. 70
   − 47

9. 21
   − 5

10. 82
   − 26

11. 312
   − 271

12. 987
   − 899

13. 500
   − 47

14. 7 385
   − 2 394

15. 2 050
   −   376

16. 26 − 12
17. 35 − 18
18. 98 − 29
19. 76 − 67
20. 312 − 97
21. 673 − 558
22. 377 − 108
23. 1 570 − 988
24. 7 803 − 5 905
25. 36 502 − 6 399
26. 19 135 − 11 236
27. 707 353 − 533 974

Use the chart for problems 28-33. Find how much greater value **A** is than value **B**.

	A	B
28.	517 inches	298 inches
29.	779 metres	488 metres
30.	2 732 pounds	976 pounds
31.	8 700 days	5 555 days
32.	12 807 litres	9 858 litres
33.	4 464 acres	1 937 acres

Solve and check each problem. The addition in parentheses is done first.

34. 87 – (35 + 19) = 33 *Ans*
35. 908 – (312 + 6 + 88)
36. 8 731 – (737 + 12 + 6 071)
37. (32 + 63 + 9) – 22
38. (503 + 7 877 + 6) – 2 033
39. (17 777 + 68 933) – 34 906
40. 700 000 – (14 393 + 5 828)

PRACTICAL APPLICATIONS

Example: A sheet metal worker fabricates ducts for a hot air heating system. One of the ducts required for the job is shown. The duct is made in six sections. Find the length of Section 4.

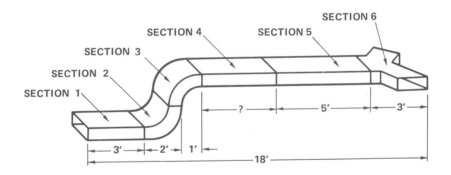

The total length and five of the six duct section lengths are given. To find the length of Section 4, add the lengths of the five sections and subtract the sum from the total length.
18 ft – (3 ft + 2 ft + 1 ft + 5 ft + 3 ft)

```
 3 ft
 2 ft
 1 ft
 5 ft
+3 ft
─────
14 ft
```

Follow the procedure for adding whole numbers.
3 ft + 2 ft + 1 ft + 5 ft + 3 ft = 14 ft

Follow the procedure for subtracting whole numbers.
18 ft – 14 ft = 4 ft

```
  18 ft
– 14 ft
──────
   4 ft Ans
```

Exercise 3-2

1. A heavy equipment operator contracts to excavate 850 cubic yards of earth for a house foundation. How much remains to be excavated after 585 cubic yards are removed?

2. A production schedule of a dress factory calls for 175 500 dresses to be completed by December 30. On December 1st 127 875 dresses are completed. How many remain to be finished?

3. An automobile mechanic determines the total bill for both labor and materials for an engine overhaul at $463. The customer pays $375 by check and the balance with cash. What amount does the customer pay by cash?

4. Five stamping machines in a manufacturing plant produce the same product. Each machine has a counter which records the number of parts produced. The table shows the counter readings for the beginning and end of one week's production.

	MACHINE 1	MACHINE 2	MACHINE 3	MACHINE 4	MACHINE 5
Counter Reading Beginning of Week	17 855	13 935	7 536	38 935	676
Counter Reading End of Week	48 951	42 007	37 881	72 302	29 275

   a. How many parts are produced during the week by each machine?
   b. What is the total weekly production?

5. A painter and decorator purchase 18 gallons of paint and 68 rolls of wallpaper for a house redecorating contract. The table lists the amount of materials that are used in each room.

	KITCHEN	LIVING ROOM	DINING ROOM	MASTER BEDROOM	SECOND BEDROOM	THIRD BEDROOM
Paint	2 gallons	4 gallons	2 gallons	3 gallons	3 gallons	2 gallons
Wallpaper	8 rolls	14 rolls	10 rolls	12 rolls	10 rolls	9 rolls

   a. Find the amount of paint remaining at the end of the job.
   b. Find the amount of wallpaper remaining at the end of the job.

6. An electrical contractor has 5 000 metres of BX cable in stock at the beginning of a wiring job. At different times during the job, electricians remove the following lengths from stock: 325 metres, 580 metres, 260 metres, and 65 metres. When the job is completed, 135 metres are left over and are returned to stock. How many metres of cable are now in stock?

7. A printer bills a customer $1 575 for an order. In printing the order, expenses are $432 for Bond paper, $287 for cover stock, $177 for envelopes, and $26 for miscellaneous materials. The customer pays the bill within 30 days and is allowed a $32 discount. How much profit does the printer make?

8. In order to make the jig shown, a machinist determines dimensions **A**, **B**, **C**, and **D**. All dimensions are in millimetres. Find **A**, **B**, **C**, and **D** in millimetres.

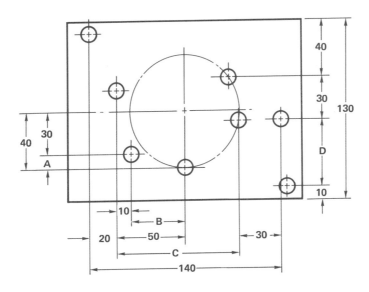

9. A small business complex is shown in the diagram. To provide parking space, a paving contractor is hired to pave the area not occupied by buildings or covered by landscaped areas. The entire parcel of land contains 41 680 square feet. How many square feet of land are paved?

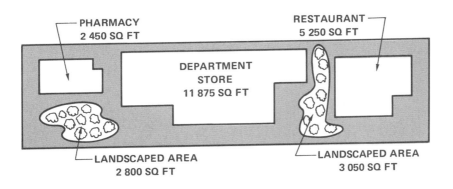

10. The table lists various kinds of flour ordered and received by a commerical baker.

	BREAD FLOUR	CAKE FLOUR	RYE FLOUR	RICE FLOUR	POTATO FLOUR	SOYBEAN FLOUR
Ordered	3 875 lb	2 000 lb	825 lb	180 lb	210 lb	85 lb
Received	3 650 lb	2 670 lb	910 lb	75 lb	165 lb	85 lb

a. Is the total amount of flour received greater or less than the total amount ordered?

b. How many pounds greater or less?

UNIT 4
Multiplication of Whole Numbers

OBJECTIVES

After studying this unit you should be able to

- Arrange and multiply whole numbers.
- Solve practical problems using multiplication.
- Solve problems by combining addition, subtraction, and multiplication.

A mason estimates the number of bricks required for a chimney. A clerk in a hardware store computes the cost of a customer's order. A secretary determines the weekly payroll of a firm. A cabinetmaker calculates the amount of plywood needed to install a store counter. A garment manufacturing supervisor determines the amounts of various materials required for a production run. These are a few of the many occupational uses of multiplication.

DEFINITIONS AND PROPERTIES OF MULTIPLICATION

Multiplication is a short method of adding equal amounts. For example, 4 times 5 (4 × 5) means 4 fives or 5 + 5 + 5 + 5.

The number to be multiplied is called the *multiplicand*. The number by which it is multiplied is called the *multiplier*. *Factors* are the numbers used in multiplying. The multiplicand and the multiplier are both factors. The result or answer of the multiplication is called the *product*. The *times sign* (×) indicates multiplication.

Numbers can be multiplied in any order. The same product is obtained regardless of the order in which the numbers are multiplied. For example, 2 × 4 × 3 may be multiplied in either of the following ways:

$$2 \times 4 \times 3 = 24 \quad \text{or} \quad 3 \times 4 \times 2 = 24$$

The numbers can also be grouped in any way and the sum is the same.

$$(2 \times 4) \times 3 \quad \text{or} \quad 2 \times (4 \times 3)$$
$$8 \times 3 = 24 \quad \quad \quad 2 \times 12 = 24$$

MULTIPLICATION TABLE

You must know the simple products from 1 times 1 to 12 times 12. If you have not completely memorized these products, practice until each multiplication fact is learned. You should be able to instantly recall every multiplication in the table.

×	1	2	3	4	5	6	7	8	9	10	11	12
1	1	2	3	4	5	6	7	8	9	10	11	12
2	2	4	6	8	10	12	14	16	18	20	22	24
3	3	6	9	12	15	18	21	24	27	30	33	36
4	4	8	12	16	20	24	28	32	36	40	44	48
5	5	10	15	20	25	30	35	40	45	50	55	60
6	6	12	18	24	30	36	42	48	54	60	66	72
7	7	14	21	28	35	42	49	56	63	70	77	84
8	8	16	24	32	40	48	56	64	72	80	88	96
9	9	18	27	36	45	54	63	72	81	90	99	108
10	10	20	30	40	50	60	70	80	90	100	110	120
11	11	22	33	44	55	66	77	88	99	110	121	132
12	12	24	36	48	60	72	84	96	108	120	132	144

PROCEDURE FOR SHORT MULTIPLICATION

Short multiplication is used to compute the product of two numbers when the multiplier contains only one digit. A problem such as 7 × 386 requires short multiplication.

Example: Multiply and check.

7 × 386

Write the multiplier under the units digit of the multiplicand.

Multiply the 7 by the units of the multiplicand.
7 × 6 = 42
Write 2 in the units position of the answer.

Multiply the 7 by the tens of the multiplicand.
7 × 8 = 56
Add the 4 tens from the product of the units.
56 + 4 = 60
Write the 0 in the tens position of the answer.

$$\begin{array}{r} 386 \\ \times 7 \\ \hline 2\,702 \text{ Ans} \end{array}$$

> Multiply the 7 by the hundreds of the multiplicand.
> 7 × 3 = 21
> Add the 6 hundreds from the product of the tens.
> 21 + 6 = 27
> Write the 7 in the hundreds position and the 2 in the thousands position.
>
> Check the answer by repeating the multiplication procedure.
>
> ```
>   3 8 6
> ×     7
> 2 7 0 2 Ck
> ```

Practice

Multiply and check.

1. 3 × 52 156
2. 8 × 692 .... 5 536
3. 4 × 1 356 .... 5 424
4. 5 × 24 066 .... 120 330
5. 7 × 136 792 .... 957 544

Exercise 4-1

Multiply and check.

1.	18 × 6	6.	546 × 3	11.	4 173 × 2	16.	12 407 × 7
2.	23 × 5	7.	606 × 4	12.	6 981 × 4	17.	26 115 × 5
3.	51 × 7	8.	934 × 9	13.	5 176 × 7	18.	99 884 × 3
4.	84 × 2	9.	418 × 6	14.	9 025 × 3	19.	33 033 × 9
5.	75 × 8	10.	775 × 5	15.	1 877 × 9	20.	54 157 × 8

21. 6 × 523
22. 3 × 1 804

23. 5 × 12 199
24. 4 × 456 900

25. 8 × 318 234
26. 9 × 2 132 512

PROCEDURE FOR LONG MULTIPLICATION

Long multiplication is used to compute the product of two numbers when the multiplier contains two or more digits. A problem such as 436 × 7 812 requires long multiplication.

Example: Multiply and check.

    436 × 7 812

Write the multiplier under the multiplicand placing digits in proper place positions.

Multiply the multiplicand by the units of the multiplier using the procedure for short muliplication.
6 × 7 812 = 46 872
Write the product starting at the units position and going from right to left.

```
        7 8 1 2
    ×     4 3 6
      4 6 8 7 2
    2 3 4 3 6
  3 1 2 4 8
  3 4 0 6 0 3 2  Ans
```

Multiply the multiplicand by the tens of the multiplier.
3 × 7 812 = 23 436
Write the product under the first product, starting at the tens position and going from right to left.

Multiply the multiplicand by the hundreds of the multiplier.
4 × 7 812 = 31 248
Write the product under the second product, starting at the hundreds position and going from right to left.

Check the answer by interchanging the multiplier with the multiplicand.

```
          4 3 6
       × 7 8 1 2
          8 7 2
        4 3 6
      3 4 8 8
    3 0 5 2
    3 4 0 6 0 3 2  Ck
```

Practice

Multiply and check.

1. 24 × 32 768
2. 24 × 896 .... 21 504
3. 12 × 4 872 .... 58 464
4. 324 × 5 723 .... 1 854 252
5. 312 × 286 .... 89 232

Exercise 4-2

Multiply and check.

1.	73 × 25	6.	384 × 44	11.	8 418 × 566	16.	19 078 × 2 146
2.	36 × 18	7.	1 512 × 67	12.	7 102 × 456	17.	513 930 × 43
3.	64 × 36	8.	8 139 × 88	13.	40 514 × 676	18.	118 125 × 219
4.	57 × 81	9.	12 737 × 79	14.	15 553 × 999	19.	327 800 × 274
5.	914 × 67	10.	3 238 × 452	15.	23 418 × 1 147	20.	405 607 × 112

21. 419 × 7 635
22. 387 × 55 676
23. 2 561 × 17 738
24. 1 176 × 62 347
25. 4 214 × 18 919

MULTIPLICATION WITH ZERO IN THE MULTIPLIER

The product of any number and zero is zero. For example, 0 × 0 = 0, 0 × 6 = 0, 0 × 8 956 = 0. When the multiplier contains zeros, the zeros must be written in the product to maintain proper place value.

Section 1 Whole Numbers

```
      674              364
    X 200            X 203
    134 800          1 092
                    72 80
                    73 892
```

Exercise 4-3

Multiply and check.

1. 85
    X 60

2. 312
    X 50

3. 943
    X 70

4. 328
    X 305

5. 746
    X 401

6. 1 798
    X 507

7. 3 307
    X 402

8. 5 625
    X 310

9. 7 100
    X 590

10. 8 041
    X 660

11. 318
    X 200

12. 8 009
    X 400

13. 900 X 6 154
14. 5 060 X 2 079
15. 700 X 6 104
16. 6 000 X 2 000

MULTIPLYING THREE OR MORE FACTORS

The multiplication of three or more numbers, two at a time, may be done in any order or in any grouping. The factors are multiplied in separate steps.

```
    7 X 5 X 2 X 3
    35    X 2 X 3
         70   X 3
              210
```

Exercise 4-4

Multiply and check.

1. 4 X 6 X 8
2. 7 X 2 X 12
3. 3 X 10 X 5
4. 2 X 5 X 12
5. 6 X 8 X 15
6. 4 X 9 X 14
7. 8 X 3 X 2 X 5
8. 5 X 4 X 6 X 3
9. 7 X 2 X 8 X 6
10. 12 X 16 X 7
11. 30 X 4 X 2 X 5
12. 10 X 5 X 40 X 3
13. 52 X 12 X 60
14. 9 X 40 X 85
15. 63 X 150 X 15 X 8

UNIT REVIEW

Exericise 4-5

Multiply and check.

1.	23 × 7	9.	8 891 × 7	17.	909 × 17	25.	4 479 × 800
2.	87 × 9	10.	8 060 × 8	18.	593 × 85	26.	7 887 × 619
3.	36 × 5	11.	85 × 63	19.	4 874 × 76	27.	7 003 × 600
4.	108 × 8	12.	94 × 77	20.	5 385 × 99	28.	31 352 × 913
5.	976 × 4	13.	68 × 12	21.	6 732 × 176	29.	87 000 × 356
6.	672 × 3	14.	43 × 70	22.	3 123 × 750	30.	6 767 × 7 804
7.	412 × 9	15.	56 × 19	23.	8 055 × 903	31.	14 932 × 8 206
8.	707 × 2	16.	818 × 38	24.	6 772 × 582	32.	9 318 × 387

33. 409 × 1 008
34. 1 763 × 8 793
35. 7 778 × 9 380
36. 3 305 × 5 617
37. 70 000 × 80 000
38. 3 × 6 × 8
39. 7 × 2 × 9
40. 5 × 8 × 3 × 7
41. 7 × 10 × 5 × 2
42. 3 × 22 × 20
43. 8 × 19 × 78
44. 13 × 27 × 66
45. 23 × 32 × 42
46. 16 × 7 × 930
47. 55 × 66 × 77
48. 9 × 803 × 43
49. 61 × 200 × 816
50. 32 × 200 × 729

28 Section 1 Whole Numbers

Use the chart for problems 51–56. Find how much greater value **A** is than value **B**.

	A	B
51.	8 × 7 pounds	5 × 6 pounds
52.	12 × 28 centimetres	13 × 14 centimetres
53.	32 × 66 hours	42 × 47 hours
54.	50 × 60 feet	50 × 50 feet
55.	9 × 125 litres	12 × 87 litres
56.	37 × 43 miles	10 × 76 miles

PRACTICAL APPLICATIONS

Practice

1. A welder finds the total length of material needed for a job. A particular job calls for 152 pieces of channel iron as shown. What total length of channel iron is needed?

   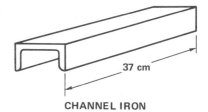
   CHANNEL IRON

   152 × 37 cm = 5 624 cm *Ans*

2. The total cost of fixtures and luminaries for an office lighting installation is found by an electrician. The following fixtures and luminaries are specified: 12 incandescent fixtures at $15 each, 22 semidirect fluorescent luminaries at $37 each, and 27 direct fluorescent luminaries at $28 each. Find the total cost.

   Total cost = (12 × $15) + (22 × $37) + (27 × $28)
   Total cost = $180 + $814 + $756
   Total cost = $1 750 *Ans*

3. A building contractor needs 8 500 feet of 2 × 4 lumber to complete projects at three different job sites. The table shown lists the inventory of 2 × 4's of various lengths at each job site. How many feet of 2 × 4's must be ordered to complete the jobs?

	NUMBER OF 2 × 4'S IN INVENTORY		
	8-foot lengths	10-foot lengths	16-foot lengths
Job Site A	120	94	52
Job Site B	36	125	67
Job Site C	82	28	110

Determine the total number of 2 × 4's of each length.

Total number of 8' lengths = 120 + 36 + 82 = 238
Total number of 10' lengths = 94 + 125 + 28 = 247
Total number of 16' lengths = 52 + 67 + 110 = 229

Determine the total number of feet of each length.

Total number of feet of 8' lengths = 238 × 8 ft = 1 904 ft
Total number of feet of 10' lengths = 247 × 10 ft = 2 470 ft
Total number of feet of 16' lengths = 229 × 16 ft = 3 664 ft

Determine the total number of feet of all lengths.

1 904 ft + 2 470 ft + 3 664 ft = 8 038 ft

Determine the number of feet of 2 × 4's to be ordered.

8 500 ft − 8 038 ft = 462 ft *Ans*

Exercise 4-6

1. An offset press feeds at the rate of 2 050 impressions per hour. How many impressions can a press operator print in 14 hours?

2. A chef estimates that an average of 150 pounds of ground beef are prepared daily. How many pounds of ground beef should be ordered for a 4 week supply? The restaurant is closed only on Mondays.

3. In drawing and dimensioning the template shown, a drafter determines distances **A, B, C, D**, and **E**. Find these distances. All dimensions are in millimetres.

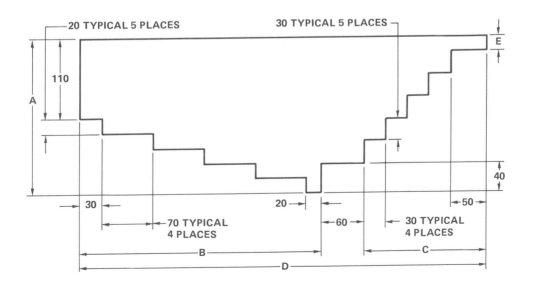

4. A wholesale distributor for small appliances orders 12 cartons of transistor radios. Each carton contains 80 radios. The wholesaler pays $8 per radio and sells each at a $6 profit. What is the total income when all the radios are sold?

5. An architectural engineering assistant determines the total weight of I beams required for a proposed building. The table lists the data used in finding the weight. Find the total weight of all I beams for the building.

   Note: The table shows the cross-section dimensions of each type of I beam. The weights given are for 1 foot of length for each type of I beam.

	20" x 7" I BEAMS WEIGHT: 80 lb/ft	18" x 6" I BEAMS WEIGHT: 55 lb/ft	12" x 5" I BEAM WEIGHT: 32 lb/ft
Number of 10-foot lengths	15	0	24
Number of 16-foot lengths	12	18	7
Number of 20-foot lengths	8	32	25
Number of 24-foot lengths	17	8	0

6. Electrical power is measured in watts. A 100-watt bulb gives more light and uses more power than a 75-watt bulb. Find the total watts used in a building which has the following lights turned on: twenty 75-watt lights, twelve 100-watt lights, four 150-watt lights, and eighteen 60-watt lights.

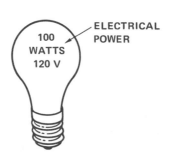

7. A gasoline dealer estimated that during the month of July (25 business days) an average of 6 500 gallons of gasoline would be sold each day. During July a total of 175 700 gallons are actually sold. How many more gallons are sold than were estimated for the month?

8. A floor tile installer measures a building to find the number of tiles needed along the edges shown. How many tiles are needed to cover the floor?

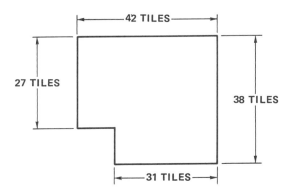

9. The size of air-conditioning equipment needed in a building depends on the number of windows and the location of the windows. The table lists the number and the amount of square feet of four different sizes of windows. The heat gain through glass in Btu/h for each square foot of glass area is shown on the table. A Btu (British Thermal Unit) is a unit of heat. Find the total heat gain (Btu/h) for the building.

	NUMBER OF WINDOWS AT EACH SIDE OF BUILDING			
Window Size (Number of square feet)	North Side 25 Btu/h/sq ft	South Side 76 Btu/h/sq ft	East Side 90 Btu/h/sq ft	West Side 99 Btu/h/sq ft
15 sq ft	8	10	4	6
24 sq ft	9	6	2	7
32 sq ft	0	4	0	3
36 sq ft	2	2	1	1

10. An excavating contractor finds that a piece of land must be drained of water before work on a job can begin. Two pumps are used to drain the water. One pump operates at the rate of 70 litres per minute for 30 minutes. The second pump operates at a rate of 90 litres per minute for 45 minutes. How many litres of water are pumped by both pumps?

11. A bookcase is produced in quantities of 1 500 by a furniture manufacturer. All pieces, except the top and back are made from 12-inch lumber. One foot of stock is allowed for cutting and waste for each bookcase. Find the total number of feet of 12-inch stock needed to manufacture the 1 500 units.

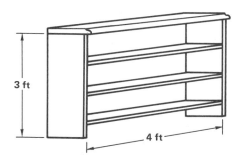

UNIT 5
Division of Whole Numbers

OBJECTIVES

After studying this unit you should be able to

- Arrange and divide whole numbers.
- Solve practical word problems using division.
- Solve problems by combining addition, subtraction, multiplication, and division.

In a grocery store unit prices are displayed. These prices are used to compare the prices for different sizes of an item. You can compare the unit price of a 25 pound bag and a 10 pound bag of the same item. You should also be able to compare items which are sold in groups. Compare the price of a product which retails at 6 for 72¢ at one store and 8 for 98¢ at another. The unit price is found by dividing.

Division is used in all occupations. The electrician must know the number of rolls of cable to order for a job. A baker determines the number of finished units made from a batch of dough. A landscaper needs to know the number of bags of lawn food required for a given area of grass. A printer determines the number of reams of paper needed for a production run of circulars.

Division is the process of finding how many times one number is contained in another. It is a short method of subtracting. Dividing 24 by 6 means to find the number of times 6 is contained in 24.

$$24 - 6 = 18$$
$$18 - 6 = 12$$
$$12 - 6 = 6$$
$$6 - 6 = 0$$

Six is subtracted 4 times from 24; therefore, 4 sixes are contained in 24.

$$24 \div 6 = 4$$

DEFINITIONS

In division, the number to be divided is called the *dividend*. The number by which the dividend is divided is called the *divisor*. The result of division is called the *quotient*. A difference left is called the *remainder*. The symbol for division is ÷. The expression 21 ÷ 7 can be written in fractional form as $\frac{21}{7}$. When written as a fraction the dividend, 21, is called the numerator and the divisor, 7, is called the denominator. The long division symbol, $\overline{)}$, is used when computing a division problem.

$$21 \div 7 \text{ is written as } 7\overline{)21}$$

ZERO AS A DIVIDEND

<u>Zero divided by a number equals zero.</u> For example, 0 ÷ 5 = 0. The fact that zero divided by a number equals zero can be shown by multiplication. The expression 0 ÷ 5 = 0 means 0 × 5 = 0. Since 0 × 5 does equal 0, it is true that 0 ÷ 5 = 0.

ZERO AS A DIVISOR

<u>Dividing by zero is impossible.</u> Students sometimes confuse division of a number by zero with division of zero by a number. It can be shown by multiplication that a number divided by zero is impossible. The expression 5 ÷ 0 = ? means that ? × 0 = 5. Since there is no real number that can be multiplied by 0 to equal 5, the division, 5 ÷ 0, is not possible. In the case of 0 ÷ 0 = ?, there is not a unique solution but there are infinite solutions. The expression 0 ÷ 0 = ? means ? × 0 = 0. Since any number times 0 equals 0, the division 0 ÷ 0 has no unique solution and is also not possible.

PROCEDURE FOR DIVIDING WHOLE NUMBERS

Write the numbers of the division problem with the divisor outside the long division symbol and the dividend within the symbol. In any division problem, the answer multiplied by the divisor plus the remainder equals the dividend.

Example: Divide and check.

4 505 ÷ 6

```
           7 5 0  R 5 Ans
       6 ) 4 5 0 5
           4 2
             3 0
             3 0
                 5
                 0
                 5
```

Write the problem with the divisor outside the long division symbol and the dividend within the symbol.

The divisor, 6, is not contained in 4, the number of thousands. The 6 will divide the 45 which is the number of hundreds. Write the 7 in the answer above the hundreds place. Subtract 42 hundreds from 45 hundreds. Write the 3 hundreds remainder in the hundreds column and add 0 tens from the dividend.

Divide 30 tens by 6. Write the 5 in the answer above the tens place. Subtract 30 tens from 30 tens. Write 5, from the dividend, in the units column.

Divide 5 by 6. Since 6 is not contained in 5, write 0 in the answer above the units place. Subtract 0 from the 5. The remainder is 5.

Check by multiplying the answer by the divisor and adding the remainder.

```
    7 5 0           4 5 0 0
  ×     6         +       5
  ─────────       ─────────────
  4 5 0 0         4 5 0 5  Ck
```

There are many practical applications which require division of whole numbers.

Example: A surveyor stakes out the boundaries of 8 building lots in the parcel of land shown. All lots have equal frontage (number of metres along the road). How many metres of frontage are in each lot?

LOT #1	LOT #2	LOT #3	LOT #4	LOT #5	LOT #6	LOT #7	LOT #8

ROAD

|← 384 m →|

Each of the lots has the same frontage. The number of metres of frontage in each lot is found by dividing 384 metres by 8.

```
           48 m Ans
       ─────────
    8 ) 384 m
        32
        ──
         64
         64
```

Practice

Divide and check.

1. $453 \div 6$ 75 R3
2. $5\ 673 \div 7$ 810 R3
3. $3\ 297 \div 8$ 412 R1
4. $43\ 682 \div 3$ 14 560 R2
5. $593\ 821 \div 9$ 65 980 R1
6. A particular factory operates two separate plants. Each plant manufactures the same three products. The table shown lists the number of each product produced at each plant during the month of February (4 weeks). Find the total average weekly production of all products at both plants.

Section 1 Whole Numbers

	NUMBER PRODUCED DURING FEBRUARY		
	Product A	Product B	Product C
Plant 1	12 850	14 760	18 925
Plant 2	10 300	13 575	16 730

Determine the total production of all products at both plants for the month of February.

12 850 + 14 760 + 18 925 + 10 300 + 13 575 + 16 730 = 87 140

Determine the average weekly production. There are 4 weeks in February. Find one equal part by dividing 87 140 by 4.

87 140 ÷ 4 = 21 785 *Ans*

Exercise 5-1

Divide and check.

1. 5)320
2. 8)256
3. 4)116
4. 7)392
5. 3)261
6. 9)405
7. 6)408
8. 2)826
9. 9)252
10. 7)378
11. 5)3 815
12. 9)1 962
13. 8)20 376
14. 4)26 356
15. 479 997 ÷ 7
16. 612 ÷ 5
17. 817 ÷ 7
18. 409 ÷ 3
19. 517 ÷ 9
20. 910 ÷ 6
21. 3 811 ÷ 2
22. $\dfrac{7\ 873}{4}$
23. $\dfrac{8\ 004}{7}$
24. $\dfrac{12\ 958}{9}$
25. $\dfrac{53\ 043}{5}$
26. $\dfrac{98\ 951}{8}$
27. $\dfrac{413\ 807}{3}$
28. $\dfrac{700\ 514}{9}$

SELECTING TRIAL QUOTIENTS

In solving long division problems, often the trial quotient selected is either too large or too small. When this occurs, another trial quotient must be selected.

Example: Divide 68 973 by 76

Write the divisor and dividend in the proper positions.

The divisor 76 is not contained in 6 or 68. Divide 689 hundreds by 76. The partial quotient is estimated as 8.

```
        8       INCORRECT
76)68 973       Trial divisor
   60 8         must be
      8 1       increased to 9.
```

Multiply: 8 × 76 = 608
Subtract: 689 − 608 = 81
The remainder, 81, is greater than the divisor, 76. The partial quotient is too small and must be increased to 9.

Note: The remainder, 81, is greater than the 76.

The problem is now correctly solved.

Divide: 689 ÷ 76
Write 9 in the partial quotient.
Multiply: 9 × 76 = 684
Subtract: 689 − 684 = 5
Bring down the 7.

```
         907  R 41 Ans
     76 ) 68 973
          68 4
             5 73
             5 32
                41
```

Divide: 57 ÷ 76 = 0
Write 0 in the partial quotient.
Bring down the 3.

Divide: 573 ÷ 76
Estimate 8 as the trial divisor.
Multiply: 8 × 76 = 608
Since 608 cannot be subtracted from 573, the trial quotient, 8, is too large and must be decreased to 7.
Multiply: 7 × 76 = 532
Subtract: 573 − 532 = 41

Check by multiplying the answer by the divisor and adding the remainder.

```
     907          68 932
   ×  76        +    41
   5 442         68 973 Ck
  63 49
  68 932
```

Maintain proper place value in division. Zeros must be shown in the quotient over their respective digits in the dividend.

Example: Divide. 24 315 006 ÷ 4 863

```
           5 000 R 6 Ans
  4 863 ) 24 315 006
          24 315
                006
```

Check: 4 863   24 315 000
        × 5 000   +   6
        24 315 000   24 315 006 Ck

Section 1 Whole Numbers

Practice

Divide and check.

1. 42 341 ÷ 68..622 R45
2. 2 362 ÷ 33..71 R19
3. 2 897 ÷ 82..35 R27
4. 7 529 ÷ 69..109 R8
5. 45 982 ÷ 26..1 768 R14

6. A painter contracts to paint the interior walls of a building. The walls are measured and found to have a total area of 16 200 square feet. Two coats of paint are required to cover the walls. Each gallon of paint covers 450 square feet. How many gallons are required for the complete job?

   Determine the number of gallons of paint required for one coat. Since one gallon covers 450 square feet, the number of gallons is determined by dividing 16 200 square feet by 450 square feet. Two coats of paint require 2 times the number of gallons used for 1 coat of paint. 72 gallons *Ans*

Exercise 5-2

Divide and check.

1. 27) 486
2. 43) 559
3. 17) 765
4. 86) 946
5. 14) 6 790
6. 32) 7 712
7. 46) 9 522
8. 73) 32 339
9. 51) 20 298
10. 620) 35 960
11. 177) 21 240
12. 806) 68 510
13. 323) 69 768
14. 515) 45 320
15. 36 650 ÷ 68
16. 95 631 ÷ 122
17. 18 661 ÷ 316
18. 30 007 ÷ 604
19. 461 079 ÷ 924
20. 507 060 ÷ 403
21. 650 000 ÷ 800
22. $\dfrac{306\ 185}{797}$
23. $\dfrac{799\ 981}{542}$
24. $\dfrac{194\ 072}{2\ 624}$
25. $\dfrac{5\ 876\ 000}{1\ 178}$
26. $\dfrac{7\ 808\ 510}{3\ 766}$
27. $\dfrac{6\ 700\ 405}{4\ 062}$
28. $\dfrac{17\ 643\ 500}{13\ 100}$

UNIT REVIEW

Exercise 5-3

Divide and check.

ONE PLACE DIVISORS

1. $4\overline{)360}$
2. $6\overline{)510}$
3. $3\overline{)285}$
4. $9\overline{)207}$
5. $8\overline{)624}$
6. $5\overline{)395}$
7. $7\overline{)487}$
8. $2\overline{)128}$
9. $9\overline{)5\,805}$
10. $6\overline{)6\,012}$
11. $2\,679 \div 3$
12. $30\,045 \div 5$
13. $389 \div 5$
14. $3\,002 \div 5$
15. $67\,393 \div 9$
16. $\frac{12\,002}{3}$
17. $\frac{87\,923}{8}$
18. $\frac{49\,006}{7}$
19. $\frac{37\,967}{4}$
20. $\frac{470\,362}{9}$

Divide and check.

MORE THAN ONE PLACE DIVISORS

21. $31\overline{)558}$
22. $25\overline{)975}$
23. $52\overline{)832}$
24. $37\overline{)814}$
25. $16\overline{)4\,848}$
26. $85\overline{)53\,270}$
27. $22\overline{)94\,380}$
28. $89\overline{)356\,712}$
29. $31\overline{)88\,660}$
30. $736\overline{)41\,216}$
31. $15\,222 \div 258$
32. $33\,948 \div 369$
33. $38\,141 \div 177$
34. $61\,910 \div 203$
35. $59\,492 \div 111$
36. $\frac{143\,268}{879}$
37. $\frac{337\,567}{427}$
38. $\frac{371\,844}{2\,817}$
39. $\frac{593\,157}{7\,053}$
40. $\frac{312\,906}{3\,981}$

Refer to the chart for problems 41-46. Find how much greater value **A** is than value **B**.

	A	B
41.	405 feet ÷ 9	266 feet ÷ 7
42.	2 496 metres ÷ 78	2 139 metres ÷ 93
43.	736 gallons ÷ 46	406 gallons ÷ 29
44.	1 856 litres ÷ 58	3 045 litres ÷ 105
45.	2 376 hours ÷ 24	6 308 hours ÷ 76
46.	6 732 acres ÷ 66	7 614 acres ÷ 81

PRACTICAL APPLICATIONS

Exercise 5-4

1. A welder fabricates 22 steel water tanks for a price of $20 570. Find the cost of each tank.

2. A garment manufacturer orders 3 000 yards of ruffling material. Each garment requires 4 yards of material. How many total garments can be produced?

3. A typist types 2 100 words in 28 minutes. How many words are typed per minute?

4. A tractor-trailer operator totals diesel fuel bills for 185 gallons of fuel used in a week. The truck travels 1 665 miles during the week. How many miles per gallon does the truck average?

5. Two sets of holes are drilled by a machinist in a piece of aluminum flat stock as shown.

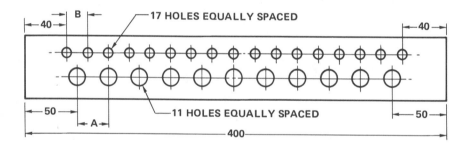

   a. Find, in millimetres, dimension **A**.
   b. Find, in millimetres, dimension **B**.

   Note: Whenever holes are arranged in a straight line, there is always one less space between the holes than the number of holes.

6. In a commercial bakery, roll dividing machines produce 16 000 dozen rolls in 8 hours. Determine the number of single rolls produced per minute.

7. A cosmetologist determines that an average of 2 ounces of liquid shampoo are required for each shampooing application. The beauty salon has 7 quarts of shampoo in stock. How many shampooing applications are made with the shampoo in stock?

   Note: One quart contains 32 ounces.

8. An ornamental iron fabricator finds the material requirements for the railing shown. How many vertical pieces of 1-inch square wrought iron are needed for this job?

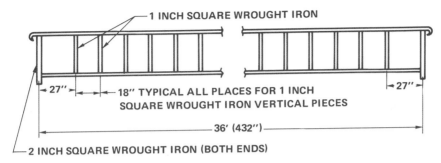

9. In estimating the time required to complete a proposed job, an electrical contractor determines that a total of 735 hours are needed. Three electricians each work 5 days per week for 7 hours per day. How many weeks are required to complete the job?

10. A chef plans the menu for a particular reception for 161 guests. The appetizer is a 6-ounce serving of tomato juice for each guest. How many 46-ounce cans of tomato juice are ordered for this reception?

11. A square sheet metal pipe is shown. A sheet metal technician takes an inventory of material needed for this job and locates five sheets of the proper gauge (thickness) stock. Each of the five sheets is 1 metre (100 centimetres) wide and 180 centimetres long. How many pipes can be made from the available stock?

Note: No allowance is made for material thickness or stretching at bends.

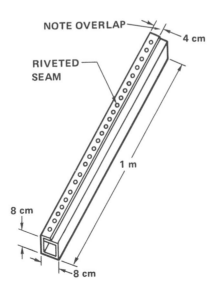

42 Section 1 Whole Numbers

12. A manufacturer produces two different products, Product A and Product B. Both products are made of aluminum, steel and copper. The table shown lists the yearly production of each product and the total amount of each metal used. Determine how much heavier one unit of Product A is than one unit of Product B.

PRODUCTION IN 1 YEAR	AMOUNT OF EACH METAL USED IN 1 YEAR		
	Aluminum	Steel	Copper
Product A: 15 800 units	19 500 kilograms	343 900 kilograms	63 200 kilograms
Product B: 18 650 units	130 550 kilograms	298 400 kilograms	55 950 kilograms

13. A landscaper contracts to prepare the soil and plant shrubbery around a group of apartment buildings. The total cost of the contract is $1 860. Find the cost per foot of length of landscaping.

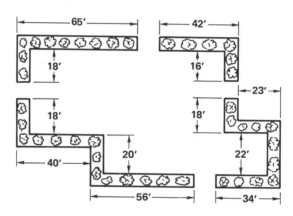

UNIT 6
Order of Operations

OBJECTIVES

After studying this unit you should be able to

- Solve arithmetic expressions by applying the proper order of operations.
- Solve problems using formulas by applying the proper order of operations.

Many occupations require the use of combined operations in solving arithmetic expressions. The arithmetic expressions are often given as formulas in occupational textbooks, manuals, and other related occupational reference materials. A *formula* uses symbols to show the relationship between quantities. The formula used in the electrical field to find the number of kilowatts of power (kW) in terms of volts (E) and amperes (I) is

$$kW = \frac{E \times I}{1\,000}$$

Following the proper order of operation is a basic requirement in solving problems involving the use of formulas.

ORDER OF OPERATIONS

A given arithmetic expression must have a unique solution. The expression $3 + 5 \times 4$ must have only one answer. The correct answer, 23, is found by using the following order of operations rules.

1. First, do all operations within grouping symbols. Grouping symbols are parentheses ( ), brackets [ ], and braces { }. The fraction bar is also a symbol used as a grouping symbol. The numerator and denominator are each considered enclosed in parentheses.

Section 1 Whole Numbers

2. Next, do multiplication and division operations in order from left to right.

3. Last, do addition and subtraction operations in order from left to right.

Example 1: Find the value of $(15 + 6) \times 3 - 28 \div 7$.

Do the work in parentheses. $(15 + 6) \times 3 - 28 \div 7$

Multiply and divide. $21 \times 3 - 28 \div 7$

Subtract. $63 - 4$

59 *Ans*

Example 2: Find the value of $\dfrac{120 - 25 \times 3}{12 + 24 \div 8} + 10$.

The fraction bar is the grouping symbol. Do all work above and below the bar first.

$$\dfrac{120 - 25 \times 3}{12 + 24 \div 8} + 10$$

$$\dfrac{120 - 75}{12 + 3} + 10$$

Divide. $\dfrac{45}{15} + 10$

Add. $3 + 10$

13 *Ans*

Practice

Find the value of each expression.

1. $4 \text{ cm} + 11 \text{ cm} - 3 \text{ cm}$ .... 12 cm
2. $4 \times 6 \div 3$ 8
3. $2 \times 60 \text{ m} + 2 \times 100 \text{ m}$ .... 320 m
4. $15 + 6 \times 3 - 28 \div 7$ 29
5. $\dfrac{(4-3) \times 6}{6} + \dfrac{10 + 2 - 4}{2}$ 5

Exercise 6-1

Perform the indicated operations.

1. $7 + 8 - 6$
2. $26 - 9 + 3$
3. $57 + 18 - 14$
4. $94 - 87 + 32 - 27$
5. $26 + 16 \div 4$
6. $(28 + 16) \div 4$
7. $\frac{72}{8} + 40$
8. $\frac{72 + 40}{8}$
9. $(72 + 40) \div 8$
10. $\frac{25 - 13 + 4}{2}$
11. $25 - 13 + \frac{4}{2}$
12. $41 + 3 \times 7 - 6$
13. $41 + 3 \times (7 - 6)$
14. $(15 \times 6) \div (3 \times 5)$
15. $(142 - 37) \div (7 \times 5)$
16. $\frac{14 + 10 \times 7}{4 + 2}$
17. $(138 - 108) \times 12 \div 6$
18. $11 \times (8 - 5) + 9 \times (2 + 7)$
19. $11 \times 8 - 5 + 9 \times 2 + 7$
20. $\frac{3 \times (29 - 6)}{23}$
21. $\frac{140}{7} + \frac{162}{54}$
22. $\frac{253 - 17 \times 3}{85 + 52 - 36}$
23. $8 \times 12 + 60 \div (9 + 3)$
24. $8 \times (12 + 60) \div (9 + 3)$
25. $(8 \times 12 + 60) \div (9 + 3)$
26. $(\frac{81}{9} + 14) \times 5$
27. $\frac{81}{9} + (14 \times 5)$
28. $30 \times (10 \times 4 - 40) \div (18 - 7)$
29. $\frac{157 - 21 \times 3}{5 - 18 \div 6} + 17$
30. $86 + \frac{27 + 8 - 5}{6} - 31$
31. $(8 \times 6 - 20) \div (7 + 21)$
32. $(8 \times 6 - 20) \div 7 + 21$
33. $\frac{576 - 16 \times 10 \times 3}{44 - 18 \times 2} - (12 - 9)$
34. $\frac{16 + 6 \times 21}{157 - 12 \times 13} - \frac{10 + 3 - 7}{46 - 4 \times 10}$

Example: An engineering technician is required to determine the size circle (diameter) needed to make the part shown. The part (a segment) must be contained within a circle. The length, dimension c, must be 20" and the height, dimension h, must be 5". The technician looks up the formula in a handbook.

$$D = \frac{c \times c + 4 \times h \times h}{4 \times h}$$

where D = diameter
  c = length of chord
  h = height of segment

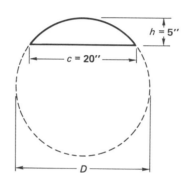

Substitute the numerical values for the variables.

$$\frac{20 \times 20 + 4 \times 5 \times 5}{4 \times 5}$$

The fraction bar is the grouping symbol. Do the work above and below the bar.

$$\frac{400 + 100}{20}$$

Divide.

$$\frac{500}{20}$$

25" *Ans*

Practice

1. The temperature of a certain solution which is read as 50°F is to be expressed in degrees Celsius.

$$°C = \frac{5 \times (°F - 32)}{9}$$

10°C *Ans*

2. A new piece of office equipment is purchased by an insurance company for $5 300. The equipment has a life expectancy (the period of usefulness) of 8 years. At the end of 8 years, it is estimated that the final or scrap value of the equipment will be $500. Find the annual depreciation.

Annual depreciation = (cost − final value) ÷ number of years of usefulness

$600 *Ans*

Exercise 6-2

1. The number of kilowatts (kW) of electrical power equals volts (E) times amperes (I) divided by 1 000.

$$kW = \frac{E \times I}{1\,000}$$

Find the number of kilowatts of power using the values given in the table.

	VOLTS (E)	AMPERES (I)	KILOWATTS (kW)
a.	110	100	
b.	115	200	
c.	220	150	
d.	230	100	
e.	220	50	

2. A retailer borrows $4 700 from a bank for 30 months. Using an installment loan table, the monthly payment on $4 000 is $158. The monthly payment on $700 is $27.65. How much total interest must the retailer pay the bank?

   Total Interest = (monthly payment on $4 000 + monthly payment on $700) × 30 − amount borrowed

3. To make an aluminum box, a sheetmetal worker finds the size of the stretchout for the box. A stretchout is a flat layout which when formed makes the box.

   Length of Stretchout = Length + 2 × height; LS = $L + 2h$
   Width of Stretchout = Width + 2 × height; WS = $W + 2h$

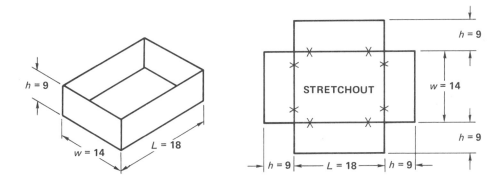

   a. Find, in centimetres, the length of the stretchout.
   b. Find, in centimetres, the width of the stretchout.

48   Section 1   Whole Numbers

4. An electrical circuit in which 3 cells are connected in series is shown.

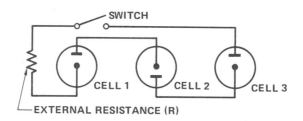

Compute the number of amperes of current in circuits a, b, and c using the values given in the table.

$$I = \frac{E \times ns}{r \times ns + R}$$ where
E = volts of one cell
ns = number of cells in circuit
r = internal resistance of one cell in ohms
R = external resistance of circuit in ohms
I = current in amperes

	E	ns	r	R	I
a.	2 volts	3 cells	1 ohm	3 ohms	
b.	5 volts	3 cells	1 ohm	2 ohms	
c.	6 volts	3 cells	1 ohm	3 ohms	

5. A comparison between Fahrenheit and Celsius scales is shown. Express the temperatures given as an equivalent degrees Fahrenheit or degrees Celsius readings.

To express degrees Fahrenheit in degrees Celsius use:

$$°C = \frac{5 \times (°F - 32)}{9}$$

To express degrees Celsius in degrees Fahrenheit use:

$$°F = \frac{9 \times °C}{5} + 32$$

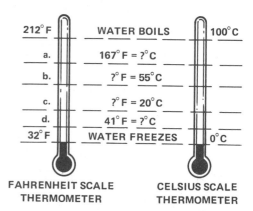

a. 167° F = ?° C
b. ?° F = 55° C
c. ?° F = 20° C
d. 41° F = ?° C

6. Automobile manufacturers sometimes use SAE rated horsepower rather than the true horsepower of an engine.

   $hp = \dfrac{2 \times D \times D \times N}{5}$ where D = cylinder bore (diameter in inches)

   N = number of cylinders

   Determine the SAE rated horsepower of a six cylinder truck engine with a cylinder bore (diameter) of 5 inches.

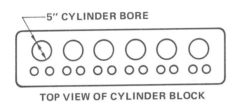

TOP VIEW OF CYLINDER BLOCK

7. Carpenters find amounts of lumber needed in board feet (bd ft). A board foot is the equivalent of a piece of lumber one foot wide, one foot long, and one inch thick.

   $Bd\ ft = \dfrac{T'' \times W'' \times L'}{12}$ where T'' = thickness in inches

   W'' = width in inches

   L' = length in feet

   Find the number of board feet in each piece of lumber shown.

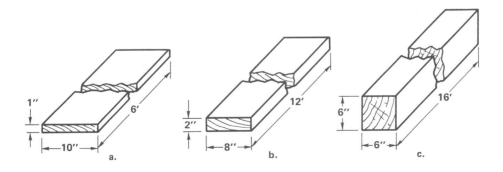

8. An electronics technician finds the total resistance (R_T) in ohms for the circuit shown.

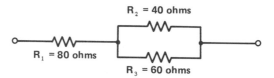

The individual resistances are represented by the symbols R_1, R_2, and R_3. Find the total resistance of the circuit.

$$R_T = R_1 + \frac{R_2 \times R_3}{R_2 + R_3}$$

SECTION 2
Common Fractions

UNIT 7
Introduction to Common Fractions

OBJECTIVES

After studying this unit you should be able to
- Determine fractional parts of quantities.
- Express fractions as equivalent fractions.
- Express fractions in lowest terms.
- Express mixed numbers as fractions.

Most measurements and calculations made on-the-job are not limited to whole numbers. Manufacturing and construction occupations require arithmetic operations using values from fractions of an inch to fractions of a mile. Food service employees prepare menus using fractions of ounces and pounds. Stock is ordered, costs are computed, and discounts are determined using fractions. Medical technicians and nurses deal with fractions when computing in the apothecaries system. Fractional arithmetic operations are necessary in the agriculture and horticulture fields in computing liquid and dry measure.

DEFINITIONS

A *fraction* is a value which shows the number of equal parts taken of a whole quantity or unit. The symbol used to indicate a fraction is the slash (/) or the bar (—).

52 Section 2 Common Fractions

The *denominator* of the fraction is the number that shows how many equal parts are in the whole quantity. The *numerator* of the fraction is the number that shows how many equal parts of the whole are taken. The numerator and denominator are called the *terms* of the fraction.

$$\frac{5}{8} \begin{matrix} \leftarrow \text{NUMERATOR} \\ \leftarrow \text{DENOMINATOR} \end{matrix}$$

FRACTIONAL PARTS

A line segment is divided into 4 equal parts.

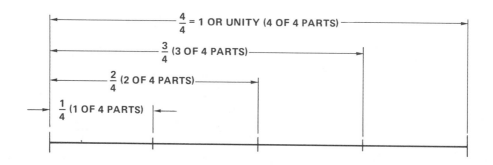

1 part = $\frac{1 \text{ part}}{4 \text{ parts}}$ = $\frac{1}{4}$ of the length of the line segment.

2 parts = $\frac{2 \text{ parts}}{4 \text{ parts}}$ = $\frac{2}{4}$ of the length of the line segment.

3 parts = $\frac{3 \text{ parts}}{4 \text{ parts}}$ = $\frac{3}{4}$ of the length of the line segment.

4 parts = $\frac{4 \text{ parts}}{4 \text{ parts}}$ = $\frac{4}{4}$ or 1.

Note: 4 parts make up the whole. ($\frac{4}{4} = 1$)

A line segment of the same length has each of the 4 equal parts divided into 8 equal parts.

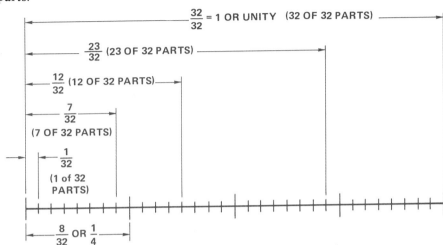

1 part   = $\frac{1}{32}$ of the whole.

7 parts   = $\frac{7}{32}$ of the whole.

12 parts = $\frac{12}{32}$ of the whole.

23 parts = $\frac{23}{32}$ of the whole.

32 parts = $\frac{32}{32}$ or 1.

Note: 8 parts = $\frac{8}{32}$ of the whole and also $\frac{1}{4}$ of the whole.

Therefore, $\frac{8}{32} = \frac{1}{4}$

Exercise 7-1

1. The total length of the line segment shown is divided into equal parts. Write the fractional part of the total length which each length, A–G, represents.

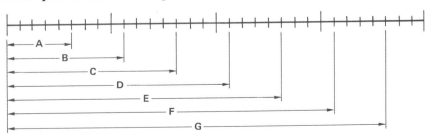

2. The total length of the line segment shown is divided into equal parts. Write the fractional part of the total length which each length, A–F, represents.

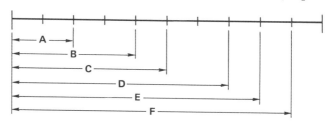

3. A riveted sheet metal plate is shown. Write the fractional part of the total number of rivets which each of the following represents.

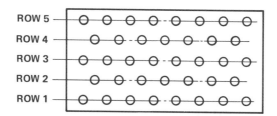

a. Row 1
b. Row 2
c. Row 2 plus row 3
d. The sum of rows 3, 4, and 5

A FRACTION AS AN INDICATED DIVISION

A fraction indicates division.

$$\frac{3}{4} \text{ means 3 is divided by 4 or } 3 \div 4$$

When performing arithmetic operations, it is sometimes helpful to write a whole number as a fraction by placing the whole number over 1. To divide by 1 does not change the value of the number.

$$5 = \frac{5}{1}$$

EQUIVALENT FRACTIONS

Equivalent fractions and equivalent units of measure use the principle of multiplying by one. Multiplication by one does not change the value. The one used is in the form of a fraction which has an equal numerator and denominator. The value of a fraction is not changed by multiplying the numerator and denominator by the same number.

> **Example:** Express $\frac{3}{8}$ as thirty-seconds.
>
> Determine what number the denominator is multiplied by to get the desired denominator. $(32 \div 8 = 4)$
>
> $\frac{3}{8} = \frac{?}{32}$
>
> Multiply the numerator and denominator by 4.
>
> $\frac{3}{8} \times \frac{4}{4} = \frac{12}{32}$ *Ans*

Practice

Express each fraction with the indicated denominator.

1. $\frac{5}{16} = \frac{?}{32}$ .... $\frac{10}{32}$
2. $\frac{7}{8} = \frac{?}{16}$ .... $\frac{14}{16}$
3. $\frac{1}{4} = \frac{?}{8}$ .... $\frac{2}{8}$
4. $\frac{3}{16} = \frac{?}{32}$ .... $\frac{6}{32}$
5. $\frac{3}{8} = \frac{?}{32}$ .... $\frac{12}{32}$

Exercise 7-2

Express each fraction as sixteenths.

1. $\frac{1}{2}$
2. $\frac{1}{4}$
3. $\frac{3}{4}$
4. $\frac{1}{8}$
5. $\frac{3}{8}$
6. $\frac{5}{8}$

Express each fraction as thirty-seconds.

7. $\frac{1}{8}$
8. $\frac{3}{4}$
9. $\frac{9}{16}$
10. $\frac{7}{8}$
11. $\frac{3}{16}$
12. $\frac{19}{16}$

Express each fraction as sixty-fourths.

13. $\frac{1}{16}$
14. $\frac{3}{8}$
15. $\frac{1}{4}$
16. $\frac{5}{32}$
17. $\frac{11}{16}$
18. $\frac{41}{32}$

Section 2 Common Fractions

Express each fraction as an equivalent fraction as indicated.

19. $\dfrac{1}{4} = \dfrac{?}{20}$

20. $\dfrac{5}{8} = \dfrac{?}{48}$

21. $\dfrac{9}{10} = \dfrac{?}{60}$

22. $\dfrac{2}{3} = \dfrac{?}{36}$

23. $\dfrac{7}{5} = \dfrac{?}{35}$

24. $\dfrac{19}{12} = \dfrac{?}{72}$

25. $\dfrac{11}{21} = \dfrac{?}{105}$

26. $\dfrac{5}{9} = \dfrac{?}{72}$

27. $\dfrac{3}{4} = \dfrac{?}{52}$

28. $\dfrac{13}{6} = \dfrac{?}{84}$

29. $\dfrac{2}{13} = \dfrac{?}{130}$

30. $\dfrac{17}{18} = \dfrac{?}{270}$

EXPRESSING FRACTIONS IN LOWEST TERMS

Multiplication and division are inverse operations. The numerator and denominator of a fraction can be divided by the same number without changing the value. A fraction is in its lowest terms when the numerator and denominator do not contain a common factor.

Arithmetic computations are usually simplified by using fractions in their lowest terms. Also, it is customary in occupations to write and speak of fractions in their lowest terms; it is part of the language of occupations. For example, a carpenter calls 6/12 foot, 1/2 foot; a machinist calls 12/16 inch, 3/4 inch; a chef calls 4/6 cup, 2/3 cup.

Example: Express $\dfrac{12}{32}$ in lowest terms.

Determine a common factor in the numerator and denominator. The numerator and the denominator can be evenly divided by 2.

$$\dfrac{12 \div 2}{32 \div 2} = \dfrac{6}{16}$$

If the fraction is not in lowest terms, find another common factor in the numerator and the denominator. Continue until the numerator and denominator have no common factor.

$$\dfrac{6 \div 2}{16 \div 2} = \dfrac{3}{8} \quad Ans$$

Practice

Express each fraction in lowest terms.

1. $\frac{2}{10}$ .... $\frac{1}{5}$
2. $\frac{70}{100}$ .... $\frac{7}{10}$
3. $\frac{16}{64}$ .... $\frac{1}{4}$
4. $\frac{16}{24}$ .... $\frac{2}{3}$
5. $\frac{8}{24}$ .... $\frac{1}{3}$

Exercise 7-3

Express each fraction as fourths.

1. $\frac{2}{8}$
2. $\frac{9}{12}$
3. $\frac{16}{64}$
4. $\frac{44}{16}$
5. $\frac{25}{20}$
6. $\frac{104}{32}$

Express each fraction as eighths.

7. $\frac{6}{16}$
8. $\frac{28}{32}$
9. $\frac{40}{64}$
10. $\frac{54}{48}$
11. $\frac{90}{80}$
12. $\frac{132}{96}$

Express each as a fraction in lowest terms.

13. $\frac{3}{9}$
14. $\frac{5}{20}$
15. $\frac{6}{30}$
16. $\frac{12}{10}$
17. $\frac{21}{14}$
18. $\frac{32}{18}$
19. $\frac{15}{75}$
20. $\frac{7}{28}$
21. $\frac{40}{16}$
22. $\frac{25}{105}$
23. $\frac{48}{64}$
24. $\frac{36}{45}$
25. $\frac{33}{88}$
26. $\frac{81}{36}$
27. $\frac{10}{55}$
28. $\frac{128}{24}$
29. $\frac{70}{26}$
30. $\frac{28}{49}$
31. $\frac{108}{24}$
32. $\frac{72}{128}$

Section 2 Common Fractions

EXPRESSING MIXED NUMBERS AS FRACTIONS

A mixed number is a whole number plus a fraction. In problem solving, it is often necessary to express a mixed number as a fraction. To express the mixed number as a fraction, find the number of fractional parts contained in the whole number then add the fractional part.

Example: Express $5\frac{1}{4}$ as a fraction.

Find the number of fractional parts contained in the whole number.

$$\frac{5}{1} \times \frac{4}{4} = \frac{20}{4}$$

Add this fraction to the fractional part of the mixed number.

$$\frac{20}{4} + \frac{1}{4} = \frac{21}{4} \; Ans$$

Practice

Express each mixed number as a fraction.

1. $1\frac{1}{2}$ .... $\frac{3}{2}$

2. $3\frac{1}{10}$ .... $\frac{31}{10}$

3. $2\frac{1}{8}$ .... $\frac{17}{8}$

4. $3\frac{3}{16}$ .... $\frac{51}{16}$

5. $4\frac{5}{8}$ .... $\frac{37}{8}$

To express a mixed number as a fraction with an indicated denominator, the mixed number is first expressed as an equivalent fraction. The fraction is then expressed as a fraction with the indicated denominator.

Example: Express the mixed number as an equivalent fraction as indicated.

$$5\frac{1}{4} = \frac{?}{12}$$

Express $5\frac{1}{4}$ as an equivalent fraction.

$$5\frac{1}{4} = \frac{21}{4}$$

Express the fraction as a fraction with the indicated denominator.

$$\frac{21}{4} = \frac{63}{12} \; Ans$$

Practice

Express each mixed number as an equivalent common fraction as indicated.

1. $1\frac{1}{2} = \frac{?}{6}$ .... $\frac{9}{6}$

2. $3\frac{1}{10} = \frac{?}{20}$ .... $\frac{62}{20}$

3. $2\frac{1}{8} = \frac{?}{40}$ .... $\frac{85}{40}$

4. $3\frac{3}{16} = \frac{?}{48}$ .... $\frac{153}{48}$

5. $4\frac{5}{8} = \frac{?}{32}$ .... $\frac{148}{32}$

EXPRESSING FRACTIONS AS MIXED NUMBERS

In solving a problem an answer may be obtained which should be expressed as a mixed number. A drafter obtains an answer of 63/32 inches. In order to make the measurement, 63/32 inches should be expressed as the mixed number 1 31/32 inches.

> **Example:** Express $\frac{60}{32}$ as a mixed number.
>
> To find how many whole units are contained, divide. Place the remainder over the denominator.
>
> $\frac{60}{32} = 1\frac{30}{32}$
>
> Express the fractional part in lowest terms.
>
> $1\frac{30}{32} = 1\frac{15}{16}$ *Ans*

Practice

Express each fraction as a mixed number.

1. $\frac{150}{100}$ .... $1\frac{1}{2}$

2. $\frac{21}{10}$ .... $2\frac{1}{10}$

3. $\frac{30}{8}$ .... $3\frac{3}{4}$

4. $\frac{18}{16}$ .... $1\frac{1}{8}$

5. $\frac{45}{32}$ .... $1\frac{13}{32}$

Exercise 7-4

Express each mixed number as a common fraction.

1. $2\frac{1}{2}$
2. $3\frac{1}{4}$
3. $5\frac{3}{4}$
4. $1\frac{5}{8}$
5. $4\frac{2}{3}$
6. $9\frac{2}{5}$
7. $8\frac{1}{3}$
8. $1\frac{3}{16}$
9. $1\frac{11}{16}$
10. $16\frac{3}{4}$
11. $150\frac{1}{2}$
12. $4\frac{9}{32}$
13. $10\frac{3}{8}$
14. $5\frac{31}{32}$
15. $63\frac{2}{3}$
16. $43\frac{4}{5}$
17. $218\frac{7}{8}$
18. $13\frac{21}{128}$
19. $302\frac{3}{4}$
20. $8\,010\frac{1}{5}$

Express each common fraction as a mixed number.

21. $\frac{3}{2}$
22. $\frac{14}{5}$
23. $\frac{10}{3}$
24. $\frac{29}{2}$
25. $\frac{63}{4}$
26. $\frac{52}{9}$
27. $\frac{47}{32}$
28. $\frac{79}{16}$
29. $\frac{96}{5}$
30. $\frac{87}{8}$
31. $\frac{133}{64}$
32. $\frac{140}{3}$
33. $\frac{103}{32}$
34. $\frac{86}{7}$
35. $\frac{217}{10}$
36. $\frac{217}{8}$
37. $\frac{451}{64}$
38. $\frac{619}{50}$
39. $\frac{412}{25}$
40. $\frac{2\,077}{32}$

Express each mixed number as an equivalent common fraction as indicated.

41. $1\frac{1}{2} = \frac{?}{8}$
42. $5\frac{3}{4} = \frac{?}{16}$
43. $7\frac{5}{8} = \frac{?}{32}$
44. $12\frac{17}{32} = \frac{?}{64}$
45. $18\frac{1}{2} = \frac{?}{10}$
46. $9\frac{3}{4} = \frac{?}{64}$
47. $15\frac{2}{3} = \frac{?}{60}$
48. $8\frac{4}{5} = \frac{?}{100}$
49. $31\frac{7}{8} = \frac{?}{32}$
50. $46\frac{9}{10} = \frac{?}{50}$
51. $24\frac{13}{16} = \frac{?}{64}$
52. $35\frac{7}{9} = \frac{?}{45}$

UNIT REVIEW
Exercise 7-5

1. The total length of the line segment is divided into equal parts. Write the fractional part of the total length which each length, A–F, represents.

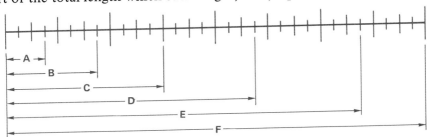

2. The pizza shown is divided into equal parts. Write the fractional part of the complete pizza which each of the following represents.

   a. 1 part   c. 7 parts   e. $\frac{1}{2}$ of 1 part

   b. 3 parts   d. 11 parts   f. $\frac{1}{4}$ of 1 part

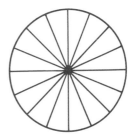

3. A patio is constructed of patio blocks all of which are the same size. Four days are required to lay the patio. Sections 1, 2, 3, and 4 are laid on the first, second, third, and fourth days respectively. What fractional part of the finished patio does each of the following represent.

   a. The number of tiles laid the first day.
   b. The number of tiles laid the second day.
   c. The number of tiles laid the third day.
   d. The number of tiles laid the fourth day.
   e. The number of tiles laid the first and second days.
   f. The number of tiles laid the second, third, and fourth days.

62 Section 2 Common Fractions

Express each of the following as thirty-seconds.

4. $\dfrac{1}{4}$
5. $\dfrac{3}{4}$
6. $\dfrac{7}{8}$
7. $\dfrac{11}{8}$
8. $\dfrac{15}{16}$
9. $\dfrac{23}{16}$

Express each of the following as sixty-fourths.

10. $\dfrac{1}{2}$
11. $\dfrac{3}{16}$
12. $\dfrac{5}{4}$
13. $\dfrac{17}{32}$
14. $\dfrac{7}{8}$
15. $\dfrac{35}{16}$

Express each of the following fractions as equivalent fractions as indicated.

16. $\dfrac{1}{2} = \dfrac{?}{8}$
17. $\dfrac{7}{12} = \dfrac{?}{36}$
18. $\dfrac{11}{15} = \dfrac{?}{60}$
19. $\dfrac{8}{7} = \dfrac{?}{35}$
20. $\dfrac{2}{9} = \dfrac{?}{72}$
21. $\dfrac{5}{3} = \dfrac{?}{54}$
22. $\dfrac{3}{16} = \dfrac{?}{256}$
23. $\dfrac{47}{51} = \dfrac{?}{357}$
24. $\dfrac{19}{6} = \dfrac{?}{132}$

Express each of the following fractions as halves.

25. $\dfrac{4}{8}$
26. $\dfrac{80}{160}$
27. $\dfrac{400}{200}$
28. $\dfrac{180}{10}$
29. $\dfrac{80}{32}$
30. $\dfrac{1\,210}{242}$

Express each of the following as a fraction in lowest terms.

31. $\dfrac{4}{10}$
32. $\dfrac{2}{8}$
33. $\dfrac{12}{8}$
34. $\dfrac{5}{35}$
35. $\dfrac{22}{44}$
36. $\dfrac{24}{6}$
37. $\dfrac{12}{28}$
38. $\dfrac{25}{150}$
39. $\dfrac{16}{64}$

Unit 7 Introduction to Common Fractions 63

40. $\dfrac{16}{128}$ 42. $\dfrac{21}{84}$ 44. $\dfrac{30}{105}$

41. $\dfrac{160}{20}$ 43. $\dfrac{140}{28}$ 45. $\dfrac{81}{18}$

Express each of the following mixed numbers as common fractions.

46. $1\dfrac{1}{2}$ 51. $9\dfrac{3}{4}$ 56. $15\dfrac{1}{4}$

47. $2\dfrac{2}{3}$ 52. $8\dfrac{1}{6}$ 57. $200\dfrac{1}{2}$

48. $1\dfrac{7}{8}$ 53. $12\dfrac{2}{5}$ 58. $3\dfrac{63}{64}$

49. $7\dfrac{4}{5}$ 54. $2\dfrac{15}{16}$ 59. $53\dfrac{3}{8}$

50. $10\dfrac{2}{3}$ 55. $1\dfrac{13}{32}$ 60. $505\dfrac{2}{3}$

Express each of the following common fractions as mixed numbers.

61. $\dfrac{5}{2}$ 66. $\dfrac{85}{4}$ 71. $\dfrac{152}{7}$

62. $\dfrac{10}{3}$ 67. $\dfrac{61}{18}$ 72. $\dfrac{167}{128}$

63. $\dfrac{12}{5}$ 68. $\dfrac{125}{122}$ 73. $\dfrac{513}{4}$

64. $\dfrac{9}{8}$ 69. $\dfrac{129}{32}$ 74. $\dfrac{319}{64}$

65. $\dfrac{23}{2}$ 70. $\dfrac{45}{13}$ 75. $\dfrac{512}{31}$

Express each of the following mixed numbers as common fractions as indicated.

76. $1\dfrac{1}{4} = \dfrac{?}{16}$ 78. $12\dfrac{2}{3} = \dfrac{?}{18}$ 80. $21\dfrac{1}{2} = \dfrac{?}{32}$

77. $7\dfrac{4}{5} = \dfrac{?}{15}$ 79. $12\dfrac{5}{8} = \dfrac{?}{64}$ 81. $26\dfrac{1}{2} = \dfrac{?}{128}$

64 Section 2 Common Fractions

82. The machined part shown is redimensioned. Express dimensions A–M in lowest terms. All dimensions are in inches.

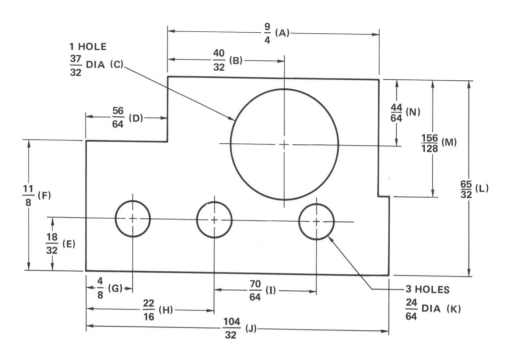

PRACTICAL APPLICATIONS

Exercise 7-6

Express all answers in lowest terms.

1. A welded support base is cut in four pieces. What fractional part of the total length does each of the four pieces represent? All dimensions are in inches.

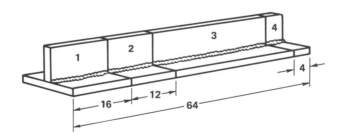

2. A parcel of land is subdivided into 5 building lots. What fractional part of the total area of the parcel of land is represented by each of the 5 lots?

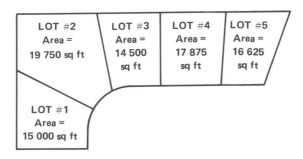

3. A retail television and appliance firm has a three-day sale on television sets. The table lists daily sales for each of three types of sets. What fractional part of the total three-day sales does each day's sales represent?

TYPE OF TELEVISION SET	FIRST DAY SALES	SECOND DAY SALES	THIRD DAY SALES
Black and White Portable	$ 7 440	$ 8 680	$11 160
Color Portable	$17 360	$16 120	$18 600
Color Console	$13 640	$12 400	$14 880

UNIT 8
Addition of Common Fractions

OBJECTIVES

After studying this unit you should be able to

- Determine lowest common denominators.
- Add fractions and mixed numbers.
- Solve practical problems using fractions and mixed numbers.

A welder determines material requirements for a certain job by adding steel plate lengths of 8 1/2", 3 1/4", and 10 3/16". In finding the costs for a garment, a garment designer adds 3/4 yard and 1 7/8 yards of woolen cloth. To straighten a particular automobile frame, an auto body specialist totals measurements of 42 3/16", 2 5/8", and 3/32". These computations require the ability to add fractions and mixed numbers.

LOWEST COMMON DENOMINATORS

Fractions cannot be added unless they have a common denominator. Common denominator means that the denominators of each of the fractions are the same, as 3/16, 7/16, and 15/16. In order to add fractions such as 1/4 and 7/8, it is necessary to determine a common denominator. The lowest common denominator is the smallest denominator which is evenly divisible by each of the denominators of the fractions being added. The lowest common denominator for the fractions 1/4 and 7/8 is 8, since 8 is the smallest number evenly divisible by both 4 and 8.

PROCEDURE FOR DETERMINING LOWEST COMMON DENOMINATORS

When it is difficult to determine the lowest common denominator, a procedure using prime factors is used. A *factor* is a number being multiplied. A *prime number* is a whole number other than 0 and 1 that is divisible only by itself and 1. A *prime factor* is a factor which is a prime number.

Unit 8 Addition of Common Fractions

To determine the lowest common denominator, first factor each of the denominators into prime factors. List each prime factor as many times as it appears in any one denominator. Multiply all of the prime factors listed.

Example 1: Find the lowest common denominator of $\frac{5}{6}$, $\frac{1}{5}$, and $\frac{3}{16}$.

Factor each of the denominators into prime factors.	6 = 2 X 3 5 = 5 (prime) 16 = 2 X 2 X 2 X 2
The prime factors 3 and 5 are each used as factors only once. List these factors. The prime factor 2 is used once for denominator 6 and 4 times for denominator 16. List 2 as a factor 4 times. If the same prime factor is used in 2 or more denominators, it is listed only for the denominator for which it is used the greatest number of times. Multiply all prime factors listed to obtain the lowest common denominator.	3 X 5 X 2 X 2 X 2 X 2 = 240 *Ans*

Example 2: Find the lowest common denominator of $\frac{9}{10}$, $\frac{8}{8}$, $\frac{4}{9}$, and $\frac{7}{15}$.

Factor each of the denominators into prime factors.	10 = 2 X 5 8 = 2 X 2 X 2 9 = 3 X 3 15 = 3 X 5
The prime factor 2 is used the greatest number of times for the denominator 8. List 2 as a factor 3 times. The prime factor 3 is used the greatest number of times for the denominator 9. List 3 as a factor 2 times. The prime factor 5 is used once for both denominator 10 and 15. List 5 as a factor only once. Multiply all prime factors listed to obtain the lowest common denominator.	2 X 2 X 2 X 3 X 3 X 5 = 360 *Ans*

Section 2 Common Fractions

Practice

Find the lowest common denominator for each set of fractions.

1. $\frac{5}{8}, \frac{1}{6}, \frac{8}{9}$ 72
2. $\frac{7}{10}, \frac{1}{5}, \frac{1}{16}$ 80
3. $\frac{4}{25}, \frac{3}{50}, \frac{1}{4}$ 100
4. $\frac{7}{20}, \frac{3}{16}, \frac{1}{4}$ 80
5. $\frac{1}{8}, \frac{1}{32}, \frac{1}{100}$ 800

Exercise 8-1

Determine the lowest common denominator for each of the following sets of fractions.

1. $\frac{1}{4}, \frac{1}{2}, \frac{3}{4}$
2. $\frac{3}{8}, \frac{1}{4}, \frac{3}{16}$
3. $\frac{1}{2}, \frac{4}{5}, \frac{7}{10}$
4. $\frac{7}{12}, \frac{2}{3}, \frac{1}{2}$
5. $\frac{1}{3}, \frac{14}{15}, \frac{2}{5}$

6. $\frac{4}{5}, \frac{9}{10}, \frac{9}{15}$
7. $\frac{5}{8}, \frac{2}{3}, \frac{7}{12}$
8. $\frac{1}{2}, \frac{7}{12}, \frac{1}{4}, \frac{7}{16}$
9. $\frac{5}{7}, \frac{1}{14}, \frac{1}{2}, \frac{3}{4}$
10. $\frac{3}{10}, \frac{1}{4}, \frac{4}{5}, \frac{5}{8}$

11. $\frac{3}{8}, \frac{1}{4}, \frac{5}{12}, \frac{1}{6}$
12. $\frac{3}{4}, \frac{1}{2}, \frac{5}{8}, \frac{5}{6}$
13. $\frac{3}{4}, \frac{7}{10}, \frac{1}{3}, \frac{8}{15}$
14. $\frac{5}{9}, \frac{3}{4}, \frac{1}{2}, \frac{1}{6}$

Using prime factors, determine the lowest common denominator for each of the following sets of fractions.

15. $\frac{7}{10}, \frac{1}{6}, \frac{8}{9}$
16. $\frac{3}{4}, \frac{4}{5}, \frac{5}{6}$
17. $\frac{1}{3}, \frac{3}{8}, \frac{9}{10}$
18. $\frac{2}{7}, \frac{1}{8}, \frac{1}{6}$
19. $\frac{7}{10}, \frac{2}{3}, \frac{3}{8}$

20. $\frac{6}{7}, \frac{9}{10}, \frac{1}{6}$
21. $\frac{13}{14}, \frac{3}{7}, \frac{5}{8}$
22. $\frac{9}{10}, \frac{1}{2}, \frac{5}{7}, \frac{2}{3}$
23. $\frac{4}{25}, \frac{7}{10}, \frac{1}{4}, \frac{5}{6}$
24. $\frac{7}{8}, \frac{3}{5}, \frac{7}{10}, \frac{1}{3}$

25. $\frac{3}{14}, \frac{7}{8}, \frac{10}{21}, \frac{3}{4}$
26. $\frac{5}{6}, \frac{1}{4}, \frac{4}{5}, \frac{9}{16}$
27. $\frac{8}{9}, \frac{1}{3}, \frac{7}{12}, \frac{3}{10}$
28. $\frac{7}{18}, \frac{9}{20}, \frac{2}{3}, \frac{3}{4}$

Comparing Values of Fractions

To compare the values of fractions with like denominators, compare the numerators. For fractions with like denominators, the larger the numerator, the larger the value of the fraction. For example, 9/16 is greater than 7/16 since 9/16 contains 9 of 16 parts and 7/16 contains only 7 of 16 parts.

To compare the values of fractions with unlike denominators, express the fractions as equivalent fractions with a common denominator and compare numerators.

Example: List the following fractions in ascending order (increasing values with smallest value first, greatest value last).

$$\frac{3''}{4}, \frac{19''}{32}, \frac{7''}{8}, \frac{41''}{64}, \frac{13''}{16}$$

Express each fraction as an equivalent fraction with a denominator of 64.

$$\frac{3''}{4} = \frac{48''}{64}$$

$$\frac{19''}{32} = \frac{38''}{64}$$

$$\frac{7''}{8} = \frac{56''}{64}$$

$$\frac{41''}{64} = \frac{41''}{64}$$

$$\frac{13''}{16} = \frac{52''}{64}$$

Compare numerators and list in ascending order.

$$\frac{19''}{32}, \frac{41''}{64}, \frac{3''}{4}, \frac{13''}{16}, \frac{7''}{8} \quad Ans$$

Observe the locations of these fractions on the enlarged inch scale.

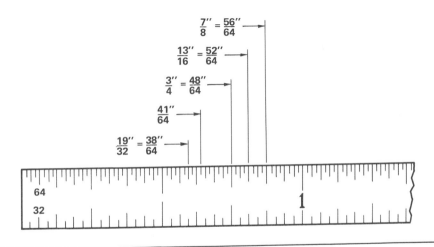

70　Section 2　Common Fractions

Practice

Arrange each set of fractions in ascending order.

1. $\frac{19}{32}, \frac{14}{20}, \frac{11}{16}$ .... $\frac{19}{32}, \frac{11}{16}, \frac{14}{20}$

2. $\frac{3}{20}, \frac{3}{12}, \frac{2}{5}$ .... $\frac{3}{20}, \frac{3}{12}, \frac{2}{5}$

3. $\frac{7}{15}, \frac{1}{6}, \frac{5}{9}$ .... $\frac{1}{6}, \frac{7}{15}, \frac{5}{9}$

4. $\frac{7}{10}, \frac{4}{15}, \frac{1}{4}$ .... $\frac{1}{4}, \frac{4}{15}, \frac{7}{10}$

5. $\frac{7}{8}, \frac{5}{12}, \frac{3}{4}$ .... $\frac{5}{12}, \frac{3}{4}, \frac{7}{8}$

Exercise 8-2

Arrange each set of fractions in ascending order (increasing values with smallest value first, greatest value last).

1. $\frac{1}{2}, \frac{1}{3}, \frac{1}{12}$
2. $\frac{3}{4}, \frac{5}{8}, \frac{1}{2}$
3. $\frac{1}{2}, \frac{3}{8}, \frac{9}{16}$
4. $\frac{9}{10}, \frac{3}{5}, \frac{1}{2}$
5. $\frac{7}{64}, \frac{9}{32}, \frac{11}{16}$
6. $\frac{7}{8}, \frac{5}{12}, \frac{3}{4}$

7. $\frac{3}{4}, \frac{9}{16}, \frac{7}{12}$
8. $\frac{3}{10}, \frac{1}{4}, \frac{7}{8}, \frac{3}{5}$
9. $\frac{3}{7}, \frac{3}{5}, \frac{8}{35}, \frac{2}{70}$
10. $\frac{7}{10}, \frac{2}{3}, \frac{4}{15}, \frac{1}{4}$
11. $\frac{11}{12}, \frac{1}{10}, \frac{1}{5}, \frac{5}{6}$
12. $\frac{13}{16}, \frac{1}{3}, \frac{1}{4}, \frac{11}{12}$

13. $\frac{1}{12}, \frac{3}{20}, \frac{2}{5}, \frac{3}{4}$
14. $\frac{7}{15}, \frac{1}{6}, \frac{5}{9}, \frac{9}{10}$
15. $\frac{3}{5}, \frac{5}{8}, \frac{1}{9}, \frac{13}{20}$
16. $\frac{1}{7}, \frac{3}{14}, \frac{5}{12}, \frac{13}{16}$
17. $\frac{8}{35}, \frac{23}{25}, \frac{1}{50}, \frac{7}{10}$
18. $\frac{4}{90}, \frac{19}{50}, \frac{43}{45}, \frac{3}{10}$

ADDING FRACTIONS

To add fractions express the fractions as equivalent fractions having the lowest common denominator. Add the numerators and write their sum over the lowest common denominator. Express the fraction in lowest terms.

Unit 8 Addition of Common Fractions

Example: Add. $\frac{5}{6} + \frac{1}{3} + \frac{7}{10} + \frac{4}{15}$

Express the fractions as equivalent fractions with 30 as the denominator.

$$\frac{5}{6} = \frac{25}{30}$$
$$\frac{1}{3} = \frac{10}{30}$$
$$\frac{7}{10} = \frac{21}{30}$$
$$+\frac{4}{15} = \frac{8}{30}$$
$$\frac{64}{30} = 2\frac{2}{15} \text{ Ans}$$

Add the numerators and write their sum over the lowest common denominator, 30.

Express the fraction in lowest terms.

Practice

A bracket is bolted to a pump assembly. In order to locate and drill bolt holes, distance A must be found. Find distance A in inches.

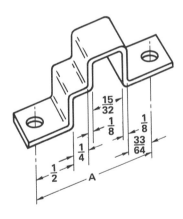

$1\frac{63}{64}$ Ans

Exercise 8-3

Add the following fractions. Express all answers in lowest terms.

1. $\frac{1}{8} + \frac{3}{8}$

2. $\frac{7}{32} + \frac{25}{32}$

3. $\frac{3}{64} + \frac{23}{64} + \frac{59}{64}$

4. $\frac{7}{8} + \frac{1}{2} + \frac{1}{4}$

72 Section 2 Common Fractions

5. $\frac{1}{2} + \frac{7}{12} + \frac{2}{3}$

6. $\frac{2}{5} + \frac{7}{10} + \frac{1}{2}$

7. $\frac{3}{4} + \frac{7}{8} + \frac{5}{16}$

8. $\frac{1}{8} + \frac{1}{4} + \frac{7}{12} + \frac{11}{12}$

9. $\frac{3}{4} + \frac{1}{8} + \frac{4}{5} + \frac{9}{10}$

10. $\frac{13}{16} + \frac{1}{4} + \frac{5}{24} + \frac{1}{2}$

11. $\frac{4}{5} + \frac{9}{35} + \frac{6}{7} + \frac{1}{5}$

12. $\frac{1}{2} + \frac{7}{12} + \frac{2}{3} + \frac{9}{16}$

13. $\frac{14}{15} + \frac{1}{3} + \frac{2}{3} + \frac{9}{10}$

14. $\frac{17}{20} + \frac{1}{5} + \frac{1}{4} + \frac{5}{12}$

15. $\frac{7}{12} + \frac{15}{24} + \frac{1}{18} + \frac{1}{4}$

16. $\frac{6}{7} + \frac{13}{14} + \frac{2}{3} + \frac{1}{2}$

17. $\frac{3}{50} + \frac{1}{10} + \frac{18}{25} + \frac{11}{35}$

18. $\frac{2}{3} + \frac{3}{8} + \frac{2}{5} + \frac{3}{20}$

19. $\frac{5}{6} + \frac{1}{9} + \frac{13}{15} + \frac{7}{20}$

20. $\frac{3}{7} + \frac{1}{12} + \frac{31}{32} + \frac{5}{6}$

ADDING FRACTIONS, MIXED NUMBERS, AND WHOLE NUMBERS

To add fractions, mixed numbers, and whole numbers, express the fractional parts of the number using a common denominator. Add the whole numbers. Add the fractions. Combine the whole number and the fraction and express in lowest terms.

Example: Add. $\frac{2}{3} + 2\frac{13}{24} + \frac{7}{12} + 15$

Express the fractional parts as equivalent fractions with 24 as the denominator.

$\frac{2}{3} = \frac{16}{24}$

$2\frac{13}{24} = 2\frac{13}{24}$

$\frac{7}{12} = \frac{14}{24}$

$+ 15 \quad = 15$

Add the whole numbers.

Add the fractions.

Combine the whole number and the fraction. Express the answer in lowest terms.

$17\frac{43}{24} = 18\frac{19}{24}$ Ans

Unit 8 Addition of Common Fractions 73

Practice

To find material requirements for a production run of oak chairs, an estimator finds the length of stock required for the chair leg. Find, in inches, the length.

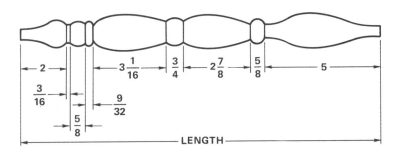

$15\frac{13''}{32}$ Ans

Exercise 8-4

Add the following values. Where necessary, express answers in lowest terms.

1. $2 + \frac{3}{4}$

2. $5\frac{1}{4} + \frac{1}{4}$

3. $7\frac{3}{4} + \frac{5}{8}$

4. $3\frac{1}{16} + 15\frac{3}{8}$

5. $17\frac{1}{10} + 14\frac{3}{5}$

6. $8\frac{1}{8} + 9\frac{5}{32}$

7. $23\frac{3}{15} + 20\frac{12}{45}$

8. $17\frac{1}{16} + 4 + 10\frac{31}{64}$

9. $9\frac{3}{8} + 1\frac{1}{2} + 15 + \frac{3}{16}$

10. $\frac{4}{15} + 3\frac{1}{3} + 7\frac{2}{3} + 18\frac{9}{10}$

11. $13\frac{1}{4} + \frac{7}{8} + 12 + 15\frac{5}{12}$

12. $39 + 1\frac{7}{12} + \frac{2}{3} + 21\frac{1}{2}$

13. $47 + 28\frac{9}{16} + 3\frac{7}{12} + 9\frac{2}{3}$

14. $14\frac{4}{5} + 107 + 5\frac{19}{35} + \frac{2}{7}$

15. $7\frac{17}{24} + 17\frac{5}{12} + 3\frac{5}{8} + 55\frac{1}{2}$

16. $18\frac{9}{10} + \frac{19}{25} + 72 + 14\frac{1}{5}$

17. $\frac{11}{20} + 66 + \frac{3}{4} + 43\frac{5}{6}$

18. $72\frac{7}{9} + 14\frac{2}{3} + 18\frac{4}{5} + \frac{1}{9}$

19. $\frac{1}{2} + \frac{9}{14} + 87 + 13\frac{3}{4}$

20. $105 + 172\frac{1}{20} + \frac{1}{15} + 66\frac{5}{9}$

21. $88\frac{1}{7} + 99\frac{1}{12} + \frac{3}{32} + 17\frac{1}{6}$

22. $\frac{29}{50} + 18\frac{17}{20} + 22\frac{19}{25} + 34$

23. $5\frac{3}{4} + 19\frac{23}{30} + \frac{3}{40} + \frac{3}{5}$

24. $\frac{1}{7} + 65\frac{15}{28} + 31 + 16\frac{3}{8}$

UNIT REVIEW

Exercise 8-5

Determine the lowest common denominators of the following sets of fractions.

1. $\dfrac{1}{2}, \dfrac{3}{4}, \dfrac{1}{8}$

2. $\dfrac{1}{4}, \dfrac{1}{16}, \dfrac{7}{8}$

3. $\dfrac{5}{12}, \dfrac{1}{3}, \dfrac{3}{8}$

4. $\dfrac{3}{5}, \dfrac{2}{3}, \dfrac{1}{10}, \dfrac{7}{15}$

5. $\dfrac{1}{3}, \dfrac{1}{4}, \dfrac{7}{8}, \dfrac{11}{12}$

6. $\dfrac{9}{16}, \dfrac{3}{4}, \dfrac{5}{12}, \dfrac{1}{2}$

7. $\dfrac{1}{2}, \dfrac{2}{14}, \dfrac{5}{7}, \dfrac{1}{4}$

8. $\dfrac{2}{5}, \dfrac{9}{10}, \dfrac{3}{4}, \dfrac{3}{8}$

Determine the lowest common denominators of the following sets of fractions by using prime factors.

9. $\dfrac{7}{16}, \dfrac{3}{5}, \dfrac{1}{6}$

10. $\dfrac{1}{15}, \dfrac{5}{6}, \dfrac{8}{9}, \dfrac{3}{10}$

11. $\dfrac{6}{7}, \dfrac{1}{10}, \dfrac{1}{2}, \dfrac{2}{3}$

12. $\dfrac{1}{4}, \dfrac{3}{5}, \dfrac{19}{20}, \dfrac{1}{6}$

13. $\dfrac{5}{12}, \dfrac{11}{14}, \dfrac{3}{16}, \dfrac{3}{7}$

14. $\dfrac{1}{50}, \dfrac{7}{75}, \dfrac{4}{45}, \dfrac{9}{150}$

Arrange each set of fractions in ascending order (increasing values with smallest value first, greatest value last).

15. $\dfrac{7}{16}, \dfrac{3}{8}, \dfrac{1}{2}$

16. $\dfrac{7}{10}, \dfrac{4}{5}, \dfrac{11}{20}$

17. $\dfrac{5}{8}, \dfrac{7}{12}, \dfrac{3}{4}$

18. $\dfrac{7}{8}, \dfrac{3}{5}, \dfrac{3}{4}, \dfrac{7}{10}$

19. $\dfrac{2}{3}, \dfrac{11}{15}, \dfrac{3}{4}, \dfrac{9}{10}$

20. $\dfrac{5}{7}, \dfrac{29}{35}, \dfrac{3}{5}, \dfrac{53}{70}$

21. $\dfrac{7}{12}, \dfrac{3}{4}, \dfrac{2}{3}, \dfrac{11}{16}$

22. $\dfrac{1}{6}, \dfrac{3}{10}, \dfrac{7}{15}, \dfrac{2}{9}$

23. $\dfrac{13}{16}, \dfrac{11}{12}, \dfrac{5}{7}, \dfrac{9}{14}$

24. $\dfrac{17}{20}, \dfrac{7}{9}, \dfrac{7}{8}, \dfrac{4}{5}$

25. $\dfrac{3}{10}, \dfrac{19}{45}, \dfrac{21}{50}, \dfrac{37}{90}$

26. $\dfrac{4}{25}, \dfrac{3}{50}, \dfrac{4}{35}, \dfrac{1}{10}$

Add the following fractions. Express all answers in lowest terms.

27. $\frac{3}{8} + \frac{5}{8}$

28. $\frac{5}{16} + \frac{7}{16}$

29. $\frac{3}{16} + \frac{1}{8} + \frac{1}{4}$

30. $\frac{9}{32} + \frac{5}{16} + \frac{3}{8}$

31. $\frac{1}{4} + \frac{5}{8} + \frac{11}{12} + \frac{1}{6}$

32. $\frac{3}{5} + \frac{5}{7} + \frac{1}{35} + \frac{3}{7}$

33. $\frac{11}{15} + \frac{2}{3} + \frac{5}{6} + \frac{7}{10}$

34. $\frac{3}{4} + \frac{2}{5} + \frac{11}{20} + \frac{5}{12}$

35. $\frac{3}{50} + \frac{7}{10} + \frac{2}{25} + \frac{13}{35}$

36. $\frac{5}{8} + \frac{3}{5} + \frac{9}{20} + \frac{2}{3}$

37. $\frac{2}{9} + \frac{4}{15} + \frac{9}{20} + \frac{5}{6}$

38. $\frac{2}{3} + \frac{1}{2} + \frac{9}{14} + \frac{1}{7}$

Add the following fractions and mixed numbers. Express all answers in lowest terms.

39. $5 + \frac{3}{8}$

40. $7\frac{1}{3} + \frac{1}{3}$

41. $9\frac{1}{2} + \frac{7}{8}$

42. $6\frac{3}{8} + 7\frac{7}{8}$

43. $24\frac{3}{10} + 31\frac{1}{5}$

44. $8\frac{9}{16} + 5 + 23\frac{17}{64}$

45. $30\frac{9}{16} + 7 + 18\frac{17}{64} + 22\frac{31}{32}$

46. $17\frac{17}{25} + 16\frac{9}{10} + 25 + 16\frac{3}{5}$

47. $81\frac{5}{24} + 19\frac{1}{2} + \frac{7}{8} + 6\frac{1}{6}$

48. $203\frac{1}{10} + 78\frac{3}{20} + 9\frac{1}{15} + 172\frac{2}{3}$

49. $14\frac{3}{4} + 6\frac{21}{30} + \frac{1}{5} + 12\frac{7}{40}$

50. $28\frac{7}{8} + 46\frac{13}{28} + \frac{3}{7} + 55$

PRACTICAL APPLICATIONS

Exercise 8-6

Solve the following problems. Express each answer in lowest terms.

1. An order of business forms which requires seven ruled columns is received by a printing shop. A printer lays out columns of the following widths: 3/4", 1 5/16", 3 1/8", 2 17/32", 1 1/2", 1 5/16", and 1 7/8". What width sheets are required for this job?

Section 2 Common Fractions

2. A cabinetmaker cuts the bottom of a dresser drawer to the dimensions shown. Find, in inches, the total length of material required.

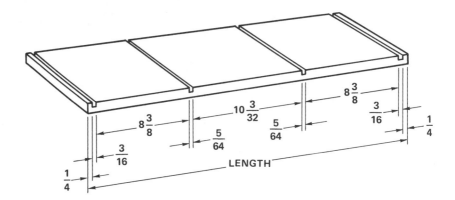

3. Determine, in inches, dimensions **A, B, C, D, E, F,** and **G** of the steel base plate shown.

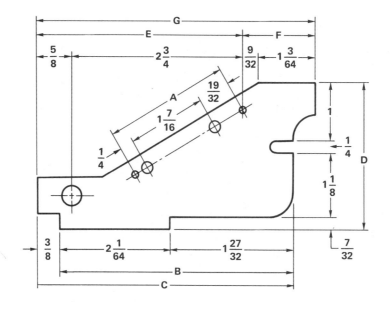

4. An appliance repair technician makes six house calls on a certain day. The times logged for the calls are 1 3/4 hours, 2/3 hour, 5/6 hour, 1 3/10 hours, 1 2/5 hours, and 4/5 hour. A total of 1 1/2 hours is spent traveling between calls and 3/4 hour is taken for lunch. If the technician begins work at 8:00 AM, at what time is the workday completed?

5. Among many computations needed in determining air conditioning specifications for a building, an environmental system technician finds window areas in square feet. The table shown lists the areas, in square feet, of different types of windows for a certain house.

	LIVING ROOM	DINING ROOM	KITCHEN	FIRST BEDROOM	SECOND BEDROOM	THIRD BEDROOM
Area of Type A Windows	$72\frac{3}{4}$	$48\frac{1}{2}$		96	$48\frac{1}{2}$	$24\frac{1}{4}$
Area of Type B Windows	$84\frac{5}{8}$	$42\frac{5}{16}$		$42\frac{5}{16}$		
Area of Type C Windows		$30\frac{1}{8}$	$60\frac{1}{4}$		$30\frac{1}{8}$	
Area of Type D Windows		$16\frac{1}{2}$	33			33

a. Find the total window area, in square feet, for each room.
b. Find the total window area, in square feet, for the entire house.

6. A plumber's piping plan consists of 6 copper pipes and 7 fittings. Both ends of each pipe are threaded 1/2 inch into the fittings. Find, in inches, the total length of the 6 pipes needed for this plan.

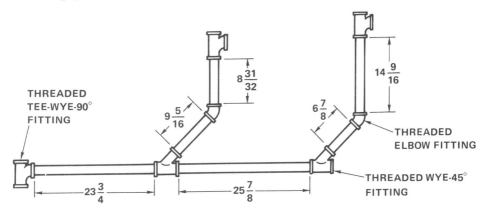

7. A furniture upholsterer takes measurements of a customer's living room set and determines the following amounts of fabric are required: 3 3/4 yards, 2/3 yard, 2 3/8 yards, 5/6 yard, and 5 7/16 yards. Find the total yards of fabric needed.

8. Four of the fence rails in the figure are each 7 3/8" wide. Find, in inches, the height of the fence post from ground level.

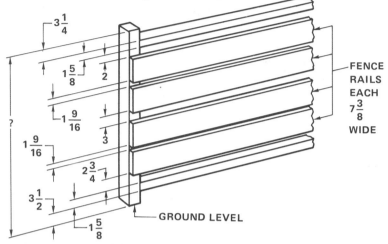

9. Three views of a machined part are drawn by a drafter. A 2 1/2" margin from each of the 4 edges of the sheet of paper is allowed.

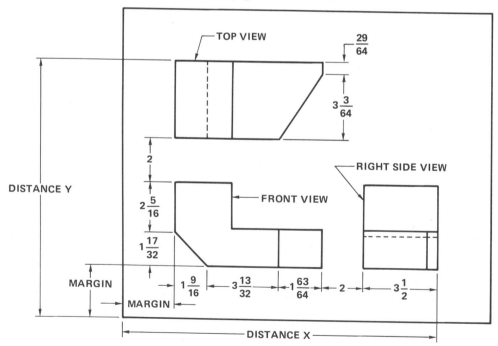

   a. Find, in inches, the distance from the left edge of the sheet to the right edge of the right side view (distance **X**).
   b. Find, in inches, the distance from the bottom of the sheet to the top edge of the top view (distance **Y**).

UNIT 9
Subtraction of Common Fractions

OBJECTIVES

After studying this unit you should be able to
- Subtract fractions from fractions.
- Subtract fractions and mixed numbers from whole numbers.
- Subtract fractions and mixed numbers from mixed numbers.
- Solve practical problems by subtraction.
- Solve practical problems by combining addition and subtraction.

While making a part from a blueprint, a machinist often finds it necessary to express blueprint dimensions as working dimensions. Subtraction of fractions and mixed numbers is used to properly position a part on a machine, to establish hole locations, and to determine depths of cuts. Subtraction of fractions and mixed numbers is used in most occupations in determining material requirements, costs, and stock sizes.

SUBTRACTING FRACTIONS FROM FRACTIONS

As in addition, fractions must have a common denominator in order to be subtracted. To subtract a fraction from a fraction, express the fractions as equivalent fractions with a common denominator. Subtract the numerators. Write their difference over the common denominator.

Example: Subtract $\frac{1}{2}$ from $\frac{5}{8}$.

Express the fractions as equivalent fractions with 8 as the denominator.

$$\frac{5}{8} = \frac{5}{8}$$
$$-\frac{1}{2} = \frac{4}{8}$$
$$\frac{1}{8} \; Ans$$

Subtract the numerators.
Write their difference 1 over the common denominator 8.

79

Practice

A metal strap is shown.

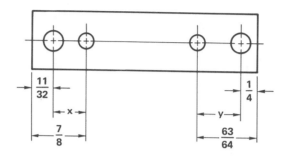

1. Find, in inches, distance **x**. $\dfrac{17''}{32}$

2. Find, in inches, distance **y**. $\dfrac{47''}{64}$

Exercise 9-1

Subtract the following fractions as indicated. Express the answers in lowest terms.

1. $\dfrac{4}{5} - \dfrac{1}{5}$
2. $\dfrac{7}{8} - \dfrac{3}{8}$
3. $\dfrac{97}{100} - \dfrac{89}{100}$
4. $\dfrac{15}{16} - \dfrac{7}{16}$
5. $\dfrac{5}{6} - \dfrac{1}{3}$
6. $\dfrac{7}{8} - \dfrac{5}{16}$
7. $\dfrac{2}{3} - \dfrac{1}{6}$
8. $\dfrac{17}{20} - \dfrac{3}{5}$
9. $\dfrac{4}{5} - \dfrac{13}{20}$
10. $\dfrac{2}{3} - \dfrac{5}{16}$
11. $\dfrac{15}{16} - \dfrac{1}{3}$
12. $\dfrac{7}{8} - \dfrac{3}{4}$
13. $\dfrac{3}{8} - \dfrac{1}{6}$
14. $\dfrac{3}{4} - \dfrac{7}{10}$
15. $\dfrac{5}{6} - \dfrac{3}{8}$
16. $\dfrac{15}{16} - \dfrac{31}{64}$
17. $\dfrac{7}{10} - \dfrac{9}{16}$
18. $\dfrac{19}{16} - \dfrac{5}{12}$
19. $\dfrac{13}{14} - \dfrac{2}{3}$
20. $\dfrac{11}{15} - \dfrac{4}{9}$
21. $\dfrac{33}{35} - \dfrac{1}{7}$
22. $\dfrac{15}{32} - \dfrac{9}{64}$
23. $\dfrac{17}{20} - \dfrac{11}{15}$
24. $\dfrac{9}{24} - \dfrac{3}{16}$

SUBTRACTING FRACTIONS AND MIXED NUMBERS FROM WHOLE NUMBERS

To subtract a fraction or a mixed number from a whole number, express the whole number as an equivalent mixed number. The fraction of the mixed number has the same denominator as the denominator of the fraction which is subtracted. Subtract the numerators and whole numbers. Combine the whole number and fraction. Express the answer in lowest terms.

Example: Subtract $\frac{13}{16}$ from 8.

Express the whole number as an equivalent mixed number.

Subtract.

$$8 = 7\frac{16}{16}$$
$$-\frac{13}{16} = \frac{13}{16}$$
$$7\frac{3}{16} \text{ Ans}$$

Practice

1. Subtract $3\frac{15}{32}$ from 15.. . . . . $11\frac{17}{32}$

2. Groove and dado joints are commonly used by furniture and cabinet makers. A groove joint runs with the grain of the wood. A dado joint runs across the grain. Find, in inches, dimensions **A** and **B**.

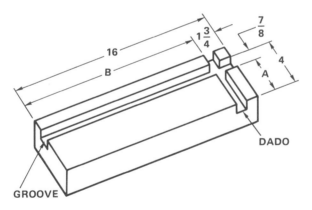

$A = 3\frac{1}{8}$ inches *Ans*

$B = 14\frac{1}{4}$ inches *Ans*

Exercise 9-2

Subtract the following values as indicated. Express the answers in lowest terms.

1. $7 - \frac{1}{2}$ 2. $5 - \frac{3}{16}$ 3. $13 - 9\frac{3}{4}$

Section 2 Common Fractions

4. $8 - 2\frac{15}{32}$
5. $18 - \frac{7}{16}$
6. $23 - \frac{3}{4}$
7. $175 - \frac{7}{10}$
8. $6 - 3\frac{7}{8}$
9. $10 - 9\frac{13}{20}$
10. $75 - 68\frac{3}{5}$
11. $109 - \frac{5}{9}$
12. $257 - \frac{29}{64}$
13. $77 - \frac{7}{45}$
14. $1 - \frac{97}{100}$
15. $312 - 310\frac{11}{32}$
16. $104 - 103\frac{38}{45}$
17. $100 - \frac{5}{124}$
18. $59 - 58\frac{37}{99}$
19. $138 - 2\frac{15}{16}$
20. $78 - \frac{75}{77}$
21. $555 - 554\frac{7}{12}$

SUBTRACTING FRACTIONS AND MIXED NUMBERS FROM MIXED NUMBERS

To subtract a fraction or a mixed number from a mixed number, the fractional part of each number must have the same denominator. Express fractions as equivalent fractions having a common denominator. When the fraction subtracted is larger than the fraction from which it is subtracted, one unit of the whole number is expressed as a fraction with the common denominator. Combine the whole number and fractions. Subtract. Express the answer in lowest terms.

Example: Subtract $\frac{5}{8}$ from $8\frac{3}{16}$.

Express the fractions as equivalent fractions with a common denominator of 16.

Since 10 is larger than 3, express one unit of the mixed number as a fraction. Combine the whole number and fractions.

Subtract.

$$8\frac{3}{16} = 8\frac{3}{16} = 7\frac{19}{16}$$
$$-\frac{5}{8} = \frac{10}{16} = \frac{10}{16}$$
$$7\frac{9}{16} \text{ Ans}$$

Practice

1. Subtract $3\frac{1}{4}$ from $7\frac{19}{64}$ .... $4\frac{3}{64}$

2. The front view of a counterbored block is shown as a section. Sectional views are often used by drafters to more clearly show the interior of a part. Imagine the block as being cut away at A-A in the top view. Find, in inches, dimension Y.

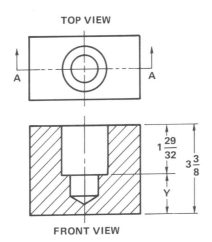

$1\dfrac{15''}{32}$ Ans

Exercise 9-3

Subtract the following values as indicated. Express the answers in lowest terms.

1. $7\dfrac{3}{8} - \dfrac{1}{8}$
2. $12\dfrac{15}{16} - \dfrac{27}{32}$
3. $9\dfrac{3}{5} - 4\dfrac{1}{5}$
4. $6\dfrac{5}{8} - 1\dfrac{1}{2}$
5. $8\dfrac{3}{10} - 3\dfrac{9}{10}$
6. $10\dfrac{13}{16} - \dfrac{3}{16}$
7. $15\dfrac{31}{32} - \dfrac{3}{4}$
8. $23\dfrac{1}{2} - 21\dfrac{1}{4}$

9. $39\dfrac{1}{32} - \dfrac{9}{16}$
10. $5\dfrac{3}{5} - 4\dfrac{27}{30}$
11. $41\dfrac{9}{20} - 29\dfrac{2}{5}$
12. $54\dfrac{19}{28} - 1\dfrac{6}{7}$
13. $21\dfrac{13}{16} - 20\dfrac{13}{16}$
14. $33\dfrac{13}{20} - 29\dfrac{5}{8}$
15. $47\dfrac{1}{4} - \dfrac{5}{7}$
16. $9\dfrac{3}{21} - 8\dfrac{19}{21}$

17. $22\dfrac{7}{12} - 13\dfrac{7}{12}$
18. $66\dfrac{3}{5} - 6\dfrac{1}{6}$
19. $79\dfrac{6}{7} - 8\dfrac{7}{8}$
20. $91\dfrac{7}{10} - \dfrac{1}{4}$
21. $101\dfrac{8}{11} - \dfrac{29}{33}$
22. $87\dfrac{1}{100} - \dfrac{2}{3}$
23. $299\dfrac{3}{32} - 298\dfrac{3}{4}$
24. $305\dfrac{4}{9} - \dfrac{1}{6}$

COMBINING ADDITION AND SUBTRACTION OF FRACTIONS AND MIXED NUMBERS

Often on-the-job computations require the combination of two or more different arithmetic operations using fractions and mixed numbers. When solving a problem which requires both addition and subtraction operations, follow the procedures learned for each operation.

Example: A sheet metal worker scribes (marks) hole locations on an aluminum sheet. The locations are scribed in the order shown: locations 1, 2, 3, and 4. Find, in inches, the distance from location 3 to location 4.

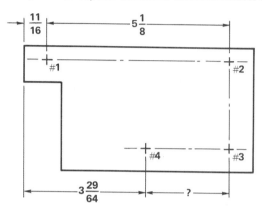

Add.

$$\frac{11''}{16} = \frac{11''}{16}$$
$$+ 5\frac{1''}{8} = 5\frac{2''}{16}$$
$$\overline{ 5\frac{13''}{16}}$$

Subtract.

$$5\frac{13''}{16} = 5\frac{52''}{64}$$
$$- 3\frac{29''}{64} = 3\frac{29''}{64}$$
$$\overline{ 2\frac{23''}{64} \text{ Ans}}$$

UNIT REVIEW

Exercise 9-4

Subtract the following fractions as indicated. Express the answers in lowest terms.

1. $\frac{7}{8} - \frac{5}{8}$

2. $\frac{15}{16} - \frac{3}{16}$

3. $\frac{39}{50} - \frac{27}{50}$

4. $\frac{5}{8} - \frac{7}{16}$

5. $\frac{1}{3} - \frac{2}{15}$

6. $\frac{13}{32} - \frac{7}{64}$

Unit 9 Subtraction of Common Fractions 85

7. $\frac{47}{100} - \frac{9}{25}$
8. $\frac{9}{10} - \frac{13}{16}$
9. $\frac{21}{35} - \frac{3}{7}$
10. $\frac{1}{2} - \frac{29}{64}$
11. $\frac{13}{24} - \frac{5}{16}$
12. $\frac{11}{20} - \frac{2}{15}$

Subtract the following values as indicated. Express the answers in lowest terms.

13. $12 - \frac{1}{4}$
14. $6 - \frac{5}{16}$
15. $18 - 10\frac{5}{8}$
16. $23 - 3\frac{9}{16}$
17. $47 - 8\frac{7}{8}$
18. $5 - \frac{59}{64}$
19. $21 - \frac{7}{32}$
20. $100 - 99\frac{77}{100}$
21. $78 - \frac{8}{25}$
22. $173 - 147\frac{35}{51}$
23. $94 - 93\frac{3}{80}$
24. $251 - 129\frac{67}{124}$

Subtract the following values as indicated. Express the answers in lowest terms.

25. $4\frac{3}{4} - \frac{1}{4}$
26. $17\frac{7}{8} - 9\frac{9}{16}$
27. $23\frac{4}{5} - 5\frac{7}{10}$
28. $34\frac{29}{32} - 30\frac{5}{8}$
29. $47\frac{3}{16} - 41\frac{9}{32}$
30. $13\frac{7}{20} - 12\frac{4}{5}$
31. $51\frac{3}{4} - \frac{6}{7}$
32. $68\frac{4}{5} - 20\frac{1}{6}$
33. $101\frac{13}{100} - \frac{2}{3}$
34. $99\frac{3}{4} - 99\frac{41}{64}$
35. $25\frac{1}{8} - \frac{7}{9}$
36. $47\frac{5}{7} - 39\frac{7}{8}$

Refer to the chart shown. For each of the problems 37-42 find how much greater value **A** is than value **B**.

	A	B
37.	$3\frac{1}{2}$ inches + $\frac{3}{4}$ inch	$1\frac{5}{16}$ inches + $1\frac{1}{8}$ inches
38.	$40\frac{1}{3}$ feet + $12\frac{1}{4}$ feet	$23\frac{1}{6}$ feet + $\frac{1}{12}$ foot
39.	$5\frac{3}{10}$ miles + $\frac{1}{3}$ mile	$\frac{5}{6}$ mile + $\frac{4}{5}$ mile
40.	$123\frac{1}{2}$ pounds + $5\frac{1}{20}$ pounds	$99\frac{1}{5}$ pounds + $26\frac{3}{4}$ pounds
41.	500 gallons + $60\frac{1}{4}$ gallons	$455\frac{1}{8}$ gallons + $\frac{1}{3}$ gallon
42.	$2\frac{1}{24}$ weeks + $\frac{2}{7}$ week	$\frac{3}{14}$ week + $\frac{1}{2}$ week

PRACTICAL APPLICATIONS

Exercise 9-5

1. Find, in inches, dimensions **A**, **B**, **C**, **D**, **E**, **F**, **G**, **H**, and **I** of the mounting plate shown.

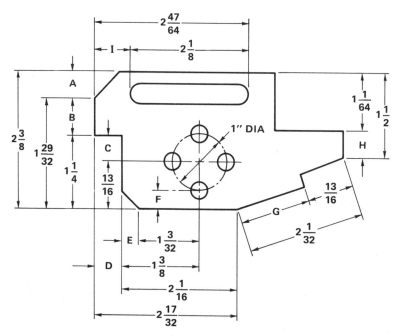

2. A wide-flange steel beam is shown. An architectural assistant determines inside dimension **C** by looking up dimensions **A** and **B** in a structural steel handbook. The table shown lists dimensions **A** and **B** for four different sizes of beams. Find dimensions **C** for each size beam.

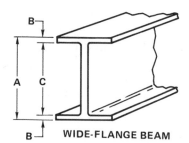

WIDE-FLANGE BEAM

	BEAM SIZES			
	6 × 4	8 × 5	8 × 7	10 × 5 $\frac{3}{4}$
A	6"	8"	8"	$9\frac{29"}{32}$
B	$\frac{9"}{32}$	$\frac{5"}{16}$	$\frac{25"}{64}$	$\frac{11"}{32}$
C	(a.)	(b.)	(c.)	(d.)

3. A baker mixes a batch of dough which weighs 200 pounds. The dough consists of 120 1/2 pounds of flour, 2 7/8 pounds of salt, 4 1/4 pounds of sugar, 1 1/3 pounds of malt, 3 2/3 pounds of shortening, and water. How many pounds of water are contained in the batch?

4. A bolt of woolen cloth is shrunk before a garment maker uses it. Before shrinking, the bolt measures 28 3/8 yards long and 3/4 yards wide. The material shrinks 2/3 yard in length and 1/36 yard in width.

   a. Find the length of the material after shrinking.
   b. Find the width of the material after shrinking.

5. A tool and die maker bores three holes in a checking gauge. The left edge and bottom edge of the gauge are the reference edges from which the hole locations are measured. Sketch and dimension the hole locations from the reference edges according to the following direction.

   Hole #1 is 1 3/32" to the right, and 1 5/8" up.
   Hole #2 is 2 1/64" to the right, and 2 3/16" up.
   Hole #3 is 3 1/4" to the right, and 3 1/2" up.

   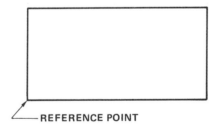
   REFERENCE POINT

   a. Find, in inches, the horizontal distance between hole #1 and hole #2.
   b. Find, in inches, the horizontal distance between hole #2 and hole #3.
   c. Find, in inches, the horizontal distance between hole #1 and hole #3.
   d. Find, in inches, the vertical distance between hole #1 and hole #2.
   e. Find, in inches, the vertical distance between hole #2 and hole #3.
   f. Find, in inches, the vertical distance between hole #1 and hole #3.

6. In estimating labor costs for a job, a bricklayer figures a total of 48 hours. The job takes longer to complete than estimated. The hours worked each day are as follows: 7 3/4, 7 1/6, 8, 9 5/6, 10 1/2, and 11 2/3. By how many hours is the job underestimated?

7. An offset duplicator operator prints forms and other printed material required by a company. In planning for three duplicating orders, the operator estimates the paper required. Four different types of bond paper are needed to print the three orders as shown in the table. Also shown is amount of paper on hand. How many total reams of all types of paper are required to complete the three orders?

Note: Paper thickness is designated by the weight of 500 sheets (1 ream). For example, the 17 x 22–Substance 16 Bond paper listed in the table means that five hundred 17 inch wide by 22 inch long sheets weigh 16 pounds.

	TYPE OF BOND PAPER			
	17 x 22 Substance 16	17 x 22 Substance 20	17 x 22 Substance 24	17 x 22 Substance 28
Paper Required for Order 1	$3\frac{3}{4}$ reams	6 reams	$\frac{1}{2}$ ream	0
Paper Required for Order 2	0	$8\frac{1}{3}$ reams	$3\frac{2}{3}$ reams	$5\frac{1}{4}$ reams
Paper Required for Order 3	$7\frac{1}{8}$ reams	0	$\frac{7}{8}$ ream	$5\frac{3}{4}$ reams
Paper on Hand	$6\frac{1}{2}$ reams	9 reams	$\frac{3}{4}$ ream	$10\frac{3}{8}$ reams

8. In finishing the interior trim of a building, a carpenter measures and saws a length of molding in three pieces. Find, in inches, the length of the third piece.

Note: 1/16 inch is allowed for each saw cut.

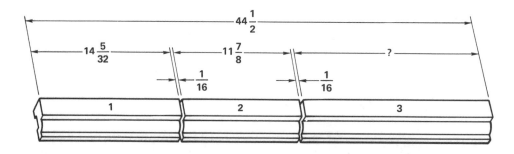

9. To reduce costs, a truck shipping firm attempts to load trucks as closely as possible to the maximum legal tonnage permitted for each truck. The shipping clerk maintains a record as shown. The record lists the tonnage loaded on five trucks at the beginning of each day for one week. The maximum legal tonnage permitted for each truck is also listed.

Truck Registration Number	Maximum Legal Tonnage Permitted	NUMBER OF TONS LOADED				
		July 8	July 9	July 10	July 11	July 12
FV-5872	12 tons	$11\frac{1}{4}$ tons	$10\frac{3}{10}$ tons	$11\frac{1}{20}$ tons	$10\frac{9}{10}$ tons	$11\frac{1}{2}$ tons
MP-2028	$10\frac{3}{4}$ tons	10 tons	$9\frac{1}{10}$ tons	$9\frac{2}{5}$ tons	$10\frac{1}{4}$ tons	$9\frac{3}{20}$ tons
GD-9395	$11\frac{1}{2}$ tons	$10\frac{9}{10}$ tons	$10\frac{3}{5}$ tons	$11\frac{1}{8}$ tons	$10\frac{7}{8}$ tons	$11\frac{1}{8}$ tons
BE-1708	10 tons	$8\frac{19}{20}$ tons	$9\frac{1}{2}$ tons	$9\frac{3}{4}$ tons	$9\frac{4}{5}$ tons	$9\frac{2}{10}$ tons

a. Find the difference between the total number of tons loaded each day and the maximum number of tons that could have been loaded each day.
b. Find the difference between the total number of tons loaded for the week and the maximum number of tons that could have been loaded for the week.

UNIT 10
Multiplication of Common Fractions

OBJECTIVES

After studying this unit you should be able to

- Multiply fractions.
- Multiply combinations of fractions, mixed numbers, and whole numbers.
- Solve practical problems by multiplication.
- Solve practical problems by combining addition, subtraction and multiplication.

A printer finds the width of a type page consisting of six 1 7/8-inch wide columns. A drygoods clerk finds the purchase price of 5 2/3 yards of fabric at $4 a yard. A welder determines the material needed for a job which requires 25 pieces of 13/16-inch long angle iron. Multiplication of fractions and mixed numbers is used for these computations.

MEANING OF MULTIPLICATION OF FRACTIONS

Just as with whole numbers, multiplication of fractions is a short method of adding equal amounts. It is important to understand the meaning of multiplication of fractions. For example, 6 × 1/4 means 1/4 is taken 6 times.

$$\frac{1}{4} + \frac{1}{4} + \frac{1}{4} + \frac{1}{4} + \frac{1}{4} + \frac{1}{4}$$

Adding the fractions gives the sum 6/4 or 1 1/2.

$$6 \times \frac{1}{4} = 1\frac{1}{2}$$

The enlarged 1-inch scale shows six 1/4-inch parts.

$$6 \times \frac{1''}{4} = 1\frac{1''}{2}$$

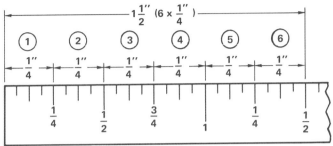

ENLARGED 1-INCH SCALE

Recall that multiplication of whole numbers can be done in any order. Multiplication of fractions can also be done in any order. The expression 6 × 1/4 is the same as 1/4 × 6. Six divided into 4 equal units is written as 1/4 of 6 or 1/4 × 6. The 6-inch scale shows six inches divided into 4 equal parts and 1 of the 4 parts is taken.

$$\frac{1}{4} \times 6'' = 1\frac{1}{2}''$$

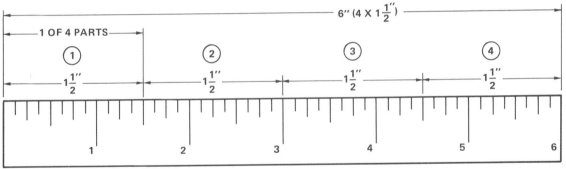

6-INCH SCALE

2 of the 4 parts equal $1\frac{1}{2}'' + 1\frac{1}{2}'' = 3''$ or $\frac{2}{4} \times 6'' = 3''$

3 of the 4 parts equal $1\frac{1}{2}'' + 1\frac{1}{2}'' + 1\frac{1}{2}'' = 4\frac{1}{2}''$ or $\frac{3}{4} \times 6'' = 4\frac{1}{2}''$

4 of the 4 parts equal $1\frac{1}{2}'' + 1\frac{1}{2}'' + 1\frac{1}{2}'' + 1\frac{1}{2}'' = 6''$ or $\frac{4}{4} \times 6'' = 6''$

Multiplication and division of fractions may result in a fractional unit of measure which is not useful. Since scales are not expressed in forty graduations, an answer of 9/40″ is not useful unless it is expressed in a unit which can be measured. To round these fractional units of measure an understanding of decimal fractions is necessary. For this section on common fractions answers will not always be expressed as practical units of measure.

Section 2 Common Fractions

The meaning of fractions multiplied by fractions and mixed numbers multiplied by fractions is the same as that which was described with a fraction times a whole number.

Examples:

1. $\frac{3}{4} \times \frac{7}{8}$ means when $\frac{7}{8}$ is divided in 4 equal parts, 3 of the 4 parts are taken.

2. $\frac{15}{16} \times 2\frac{3}{32}$ means when $2\frac{3}{32}$ is divided in 16 equal parts, 15 of the 16 parts are taken.

Exercise 10-1

For each of the following statements, insert the proper numerical value for a and b.

1. $\frac{2}{5} \times \frac{3}{4}$ means when $\frac{3}{4}$ is divided in a equal parts, b of the a parts are taken.

2. $\frac{17}{32} \times \frac{7}{8}$ means when $\frac{7}{8}$ is divided in a equals parts, b of the a parts are taken.

3. $\frac{5}{3} \times \frac{1}{2}$ means when $\frac{1}{2}$ is divided in a equal parts, b of the a parts are taken.

4. $\frac{9}{16} \times 3\frac{15}{32}$ means when $3\frac{15}{32}$ is divided in a equal parts, b of the a parts are taken.

5. $\frac{127}{64} \times 10\frac{9}{10}$ means when $10\frac{9}{10}$ is divided in a equal parts, b of a parts are taken.

MULTIPLYING FRACTIONS

To multiply two or more fractions, multiply the numerators. Multiply the denominators. Write as a fraction with the product of the numerators over the product of the denominators. Express the answer in lowest terms.

Example: Multiply. $\frac{2}{3} \times \frac{4}{5}$

Multiply the numerators.
Multiply the denominators.
Write as a fraction.

$\frac{2}{3} \times \frac{4}{5} = \frac{8}{15}$ *Ans*

Unit 10 Multiplication of Common Fractions

Practice

1. Multiply as indicated.

   $\frac{1}{2} \times \frac{3}{4} \times \frac{5}{6}$ .... $\frac{5}{16}$

2. A hole is to be drilled in a block to a depth of 3/4 of the thickness of the block. Find the depth of the hole.

   $\frac{45''}{64}$ Ans

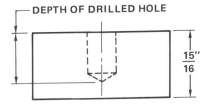

Exercise 10-2

Multiply the following fractions as indicated. Express the answers in lowest terms.

1. $\frac{1}{2} \times \frac{1}{2}$
2. $\frac{1}{4} \times \frac{1}{2}$
3. $\frac{1}{4} \times \frac{3}{4}$
4. $\frac{1}{3} \times \frac{2}{3}$
5. $\frac{4}{5} \times \frac{7}{8}$
6. $\frac{3}{16} \times \frac{3}{8}$
7. $\frac{5}{8} \times \frac{5}{8}$
8. $\frac{9}{10} \times \frac{7}{10}$
9. $\frac{5}{6} \times \frac{11}{12}$
10. $\frac{1}{6} \times \frac{2}{5}$
11. $\frac{19}{20} \times \frac{3}{5}$
12. $\frac{8}{9} \times \frac{6}{7}$
13. $\frac{3}{4} \times \frac{19}{32}$
14. $\frac{4}{7} \times \frac{7}{15}$
15. $\frac{3}{4} \times \frac{2}{3} \times \frac{9}{10}$
16. $\frac{4}{5} \times \frac{7}{8} \times \frac{5}{8}$
17. $\frac{5}{6} \times \frac{17}{20} \times \frac{1}{15}$
18. $\frac{11}{12} \times \frac{5}{6} \times \frac{3}{20}$
19. $\frac{8}{9} \times \frac{18}{21} \times \frac{3}{8} \times \frac{1}{9}$
20. $\frac{3}{4} \times \frac{12}{25} \times \frac{10}{11} \times \frac{22}{25}$
21. $\frac{9}{32} \times \frac{4}{9} \times \frac{11}{16} \times \frac{4}{33}$

MULTIPLYING ANY COMBINATION OF FRACTIONS, MIXED NUMBERS, AND WHOLE NUMBERS

To multiply any combination of fractions, mixed numbers, and whole numbers, write the mixed numbers as fractions. Write whole numbers over the denominator 1. Multiply numerators. Multiply denominators. Express the answer in lowest terms.

94 Section 2 Common Fractions

> **Example:** Multiply as indicated.
>
> $$4 \times \frac{7}{8}$$
>
> Write the whole number, 4, over 1.
>
> Multiply numerators.
> Multiply denominators.
>
> Express the answer in lowest terms.
>
> $$\frac{4}{1} \times \frac{7}{8} = \frac{28}{8} \text{ or } 3\frac{1}{2} \text{ Ans}$$

Practice

Multiply.

1. $\frac{6}{7} \times 5\frac{1}{2}$ .... $4\frac{5}{7}$

2. $2\frac{3}{5} \times 6\frac{7}{10}$ .... $17\frac{21}{50}$

3. $\frac{3}{4} \times 12 \times 4\frac{5}{8}$ .... $41\frac{5}{8}$

DIVIDING BY COMMON FACTORS

Problems involving multiplication of fractions are generally solved more quickly and easily if a numerator and a denominator are divided by any common factors before the fractions are multiplied.

> **Examples:**
>
> 1. Multiply $\frac{3}{4} \times \frac{8}{9}$.
>
>   The factor 3 is common to both the numerator 3 and the denominator 9. Divide 3 and 9 by 3.
>
>   The factor 4 is common to both the denominator 4 and the numerator 8. Divide 4 and 8 by 4.
>
>   Multiply numerators.
>   Multiply denominators.
>
> $$\frac{\overset{1}{\cancel{3}}}{\cancel{4}} \times \frac{\overset{2}{\cancel{8}}}{\cancel{9}} = \frac{2}{3} \text{ Ans.}$$

2. Multiply $2\frac{2}{5} \times 6\frac{7}{8}$.

   Express the mixed numbers as fractions.
   Divide 5 and 55 by 5.
   Divide 12 and 8 by 4.
   Multiply numerators.
   Multiply denominators.
   Express the answer in lowest terms.

$$\frac{\cancel{12}^3}{\cancel{5}_1} \times \frac{\cancel{55}^{11}}{\cancel{8}_2} = \frac{33}{2} = 16\frac{1}{2} \; Ans$$

Practice

Multiply.

1. $\frac{4}{7} \times \frac{5}{18} \times \frac{14}{15}$ .... $\frac{4}{27}$

2. $\frac{5}{14} \times \frac{8}{9} \times \frac{7}{10}$ .... $\frac{2}{9}$

3. $2\frac{2}{3} \times 12 \times \frac{5}{16}$ 10

4. A floor refinisher removes the finish and applies a new finish to wood floors. In order to quote a price to a customer, the number of square feet (area) of a floor must be found. The length and width of a floor are measured. The area of this floor equals the length times the width. Determine the number of square feet to be refinished.

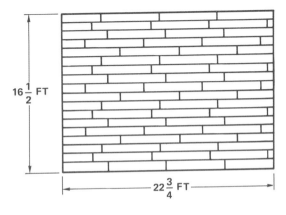

$375\frac{3}{8}$ sq ft *Ans*

96 Section 2 Common Fractions

5. A machine produces 180 pieces per hour. An average of 1/50 of the pieces are defective and must be reworked or scrapped. How many defective pieces are produced in 12 1/2 hours?

45 *Ans*

Exercise 10-3
Multiply the following values as indicated. Express the answers in lowest terms.

1. $\frac{1}{2} \times 5$
2. $\frac{3}{4} \times 12$
3. $\frac{7}{8} \times 11$
4. $15 \times \frac{3}{5}$
5. $10 \times \frac{31}{32}$
6. $\frac{1}{9} \times 2$
7. $\frac{5}{9} \times 1\frac{1}{2}$
8. $\frac{3}{4} \times 3\frac{2}{3}$
9. $\frac{7}{8} \times 3\frac{1}{5}$
10. $\frac{1}{2} \times 10\frac{3}{8}$
11. $\frac{9}{16} \times 3\frac{5}{9}$
12. $12\frac{2}{3} \times \frac{6}{19}$
13. $7\frac{11}{32} \times \frac{8}{15}$
14. $12\frac{2}{5} \times \frac{3}{4}$
15. $2\frac{2}{5} \times 1\frac{1}{3}$
16. $5\frac{1}{4} \times 1\frac{5}{7}$
17. $4\frac{5}{16} \times 2\frac{2}{3}$
18. $10\frac{5}{8} \times 7\frac{1}{2}$
19. $83\frac{3}{4} \times 2\frac{2}{15}$
20. $\frac{3}{4} \times 12 \times 1\frac{5}{8}$
21. $10 \times \frac{7}{32} \times 3\frac{1}{7}$
22. $4\frac{4}{5} \times 25 \times 2\frac{1}{16}$
23. $13 \times \frac{23}{26} \times 2\frac{1}{46}$
24. $7\frac{1}{2} \times 80 \times \frac{1}{3}$

COMBINING ADDITION, SUBTRACTION, AND MULTIPLICATION OF FRACTIONS, MIXED NUMBERS, AND WHOLE NUMBERS

When solving a problem which requires two or more different operations, think the problem through to determine the steps used in its solution. Then follow the procedures for each operation.

Example:

Thirty welded pipe supports are fabricated to the dimensions shown. A cut-off and waste allowance of 3/4" is made for each piece. A total of 24'-3 3/8" (291 3/8") of channel iron of the required size are in stock. How many feet of channel iron are ordered for the complete job?

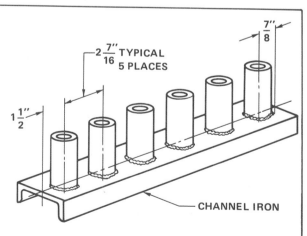

Find the length of channel iron required for one support.

$$5 \times 2\frac{7''}{16} = \frac{5}{1} \times \frac{39''}{16} = \frac{195''}{16} = 12\frac{3''}{16}$$

$$1\frac{1''}{2} = 1\frac{8''}{16}$$

$$12\frac{3''}{16} = 12\frac{3''}{16}$$

$$+ \frac{7''}{8} = \frac{14''}{16}$$

$$13\frac{25''}{16} = 14\frac{9''}{16}$$

Find the length of one support including the cut-off and waste allowance.

$$14\frac{9''}{16} = 14\frac{9''}{16}$$

$$+ \frac{3''}{4} = \frac{12''}{16}$$

$$14\frac{21''}{16} = 15\frac{5''}{16}$$

Find the length of 30 supports.

$$30 \times 15\frac{5''}{16} = \frac{30}{1} \times \frac{245''}{16} = \frac{7\,350''}{16} = 459\frac{3''}{8}$$

Find the amount of channel iron ordered.

$$459\frac{3''}{8}$$

$$-291\frac{3''}{8}$$

$$168''$$

Express the answer in feet.

$$168 \div 12 = 14$$

$$168'' = 14' \; Ans$$

UNIT REVIEW

Exercise 10-4

Multiply the following fractions as indicated. Express the answers in lowest terms.

1. $\frac{1}{4} \times \frac{1}{2}$

2. $\frac{1}{8} \times \frac{3}{4}$

3. $\frac{2}{3} \times \frac{7}{8}$

4. $\frac{9}{16} \times \frac{5}{6}$

Section 2 Common Fractions

5. $\frac{8}{9} \times \frac{3}{32}$

6. $\frac{1}{4} \times \frac{2}{9} \times \frac{3}{16}$

7. $\frac{11}{20} \times \frac{5}{16} \times \frac{4}{5}$

8. $\frac{5}{6} \times \frac{3}{25} \times \frac{5}{32}$

9. $\frac{9}{16} \times \frac{2}{3} \times \frac{1}{4} \times \frac{1}{2}$

10. $\frac{18}{25} \times \frac{5}{9} \times \frac{6}{25} \times \frac{1}{4}$

11. $\frac{7}{8} \times \frac{4}{5} \times \frac{5}{8} \times \frac{2}{3}$

12. $\frac{4}{9} \times \frac{3}{16} \times \frac{4}{5} \times \frac{5}{6}$

Multiply the following values as indicated. Express the answers in lowest terms.

13. $\frac{3}{8} \times 16$

14. $20 \times \frac{4}{5}$

15. $\frac{1}{7} \times 28$

16. $\frac{15}{16} \times 5\frac{1}{3}$

17. $10\frac{2}{3} \times 9\frac{1}{4}$

18. $7\frac{5}{16} \times 3\frac{1}{3}$

19. $\frac{3}{8} \times 16 \times 2\frac{3}{4}$

20. $12 \times \frac{9}{32} \times 2\frac{2}{7}$

21. $5\frac{1}{3} \times \frac{9}{16} \times 10\frac{1}{2}$

22. $6\frac{4}{5} \times 30 \times 3\frac{1}{8}$

23. $20 \times \frac{15}{64} \times \frac{4}{25}$

24. $8\frac{3}{4} \times \frac{7}{10} \times \frac{2}{35}$

Refer to the chart shown. For each of the problems 25-30, find how much greater value **A** is than value **B**.

	A	B
25.	$\frac{2}{3} \times \frac{3}{4}$ pound	$\frac{1}{2} \times \frac{5}{8}$ pound
26.	$\frac{15}{16} \times 32$ inches	$\frac{5}{32} \times 74$ inches
27.	$\frac{1}{5} \times 65\frac{3}{4}$ hours	$\frac{3}{10} \times 17\frac{2}{3}$ hours
28.	$20\frac{7}{12} \times 23$ feet	$37 \times 8\frac{1}{2}$ feet
29.	$19\frac{15}{16} \times 35\frac{1}{8}$ gallons	$3\frac{1}{5} \times 88\frac{5}{32}$ gallons
30.	$\frac{3}{10} \times 73\frac{9}{10}$ miles	$\frac{9}{10} \times 14\frac{17}{20}$ miles

PRACTICAL APPLICATIONS

Exercise 10-5

1. The Unified Thread may have either a flat or rounded crest or root. If the sides of the Unified Thread are extended a sharp V-thread is formed. **H** is the height of a sharp V-thread. The pitch, **P**, is the distance between two adjacent threads.

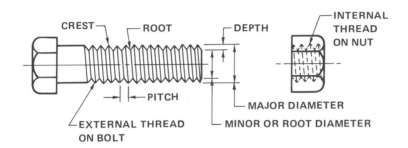

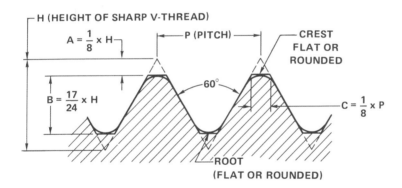

Find dimensions **A**, **B**, and **C** as indicated.

a. $H = \frac{5''}{16}$, $A = ?$, $B = ?$ f. $P = \frac{1''}{4}$, $C = ?$

b. $H = \frac{3''}{8}$, $A = ?$, $B = ?$ g. $P = \frac{1''}{6}$, $C = ?$

c. $H = \frac{15''}{64}$, $A = ?$, $B = ?$ h. $P = \frac{1''}{20}$, $C = ?$

d. $H = \frac{1''}{2}$, $A = ?$, $B = ?$ i. $P = \frac{1''}{28}$, $C = ?$

e. $H = \frac{3''}{4}$, $A = ?$, $B = ?$ j. $P = \frac{1''}{32}$, $C = ?$

2. The recipe for chicken salad makes 4 servings. A chef finds the amount of each ingredient for a serving of 25 on one day and a serving of 42 on the next. Find the amount of each ingredient needed for 25 servings and 42 servings.

   Hint: For the 25 servings, multiply each ingredient by 25/4 or 6 1/4.

		AMOUNT OF EACH INGREDIENT REQUIRED		
		4 Servings	25 Servings	42 Servings
a.	Diced cooked chicken	$1\frac{3}{4}$ cups	? cups	? cups
b.	Sliced celery	$1\frac{1}{4}$ cups	? cups	? cups
c.	Lemon juice	1 teaspoon	? teaspoons	? teaspoons
d.	Chopped green onions	2	?	?
e.	Salt	$\frac{1}{2}$ teaspoon	? teaspoons	? teaspoons
f.	Paprika	$\frac{1}{8}$ teaspoon	? teaspoons	? teaspoons
g.	Medium size avocado	1	?	?
h.	Cashew nuts	$\frac{1}{2}$ cup	? cups	? cups
i.	Mayonnaise	$\frac{1}{3}$ cup	? cups	? cups

3. An interior decorator applies the following steps in computing the width of non-draw sheer window draperies.

   - Measures the window width.
   - Triples the window width measurement.
   - Allows 4 inches (1/3 foot) for each side hem (2 required).

   a. Find, in feet, the width of drapery fabric needed for three 6 1/4-foot wide windows.
   b. Find, in feet, the width of drapery fabric needed for two 3 1/2-foot wide windows.
   c. Find, in feet, the total width needed for all the windows.

4. A retail clothing store has an end-of-the-season clearance sale on summer clothing. The amount of price markdown depends on the particular piece of merchandise. A customer purchases the items shown in the table. A salesclerk finds the cost of all markdown items. What is the customer charged for the clothing? Note: 1/4 off the list price means 1/4 of the listed price is deducted or the markdown price is 3/4 of the list price.

ITEM	LIST PRICE	MARKDOWN FROM LIST PRICE	NUMBER PURCHASED
#1	$7	$\frac{1}{4}$ off	3
#2	$12	$\frac{1}{3}$ off	2
#3	$15	$\frac{1}{4}$ off	3
#4	$23	$\frac{1}{2}$ off	1

5. A carpenter contracts to remodel an older home. The present stairway shown is removed and replaced. To order materials, the total run and rise of the present stairway are found.

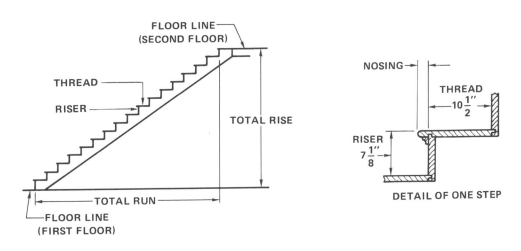

a. Find the total rise.
b. Find the total run.

102 Section 2 Common Fractions

6. A sheet metal technician is required to cut twenty-five 3 9/16" lengths of band iron allowing 3/32" waste for each cut. The pieces are cut from a strip of band iron which is 121 3/4" long. How much stock is left after all pieces are cut?

7. A printer selects type for a book which averages 12 1/2 words per line and 42 lines per page. The book has five chapters. The number of pages in each chapter are as follows: 30 1/2, 37, 28 2/3, 40 1/4, and 43 1/3. How many estimated words are in the book?

8. An elevation of a brick fireplace is shown. The bricks are 2 1/2" x 3 7/8" x 8 1/4". All mortar joints are 3/8" thick.

   Note: The number of mortared joints is one less than the number of bricks.

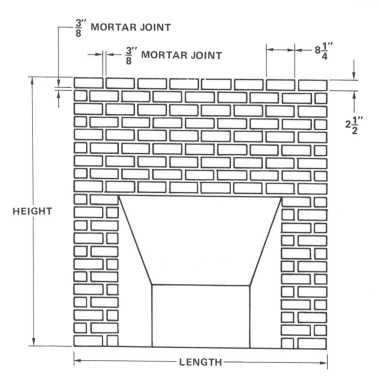

   a. Find the length of the fireplace.
   b. Find the height of the fireplace.

UNIT 11
Division of Common Fractions

OBJECTIVES

After studying this unit you should be able to

- Divide fractions.
- Divide any combination of fractions, mixed numbers, and whole numbers.
- Solve practical problems by division.
- Solve practical problems by combining addition, subtraction, multiplication, and division.

Division of fractions and mixed numbers is used by a chef to determine the number of servings that can be prepared from a given quantity of food. A painter and decorator finds the number of gallons of paint needed for a job. A cosmetologist finds the number of applications that can be obtained from a certain amount of hair rinse solution. A cabinetmaker determines shelving spacing for a counter installation.

MEANING OF DIVISION OF FRACTIONS

As with division of whole numbers, division of fractions is a short method of subtracting. Dividing 1/2 by 1/8 means to find the number of times 1/8 is contained in 1/2.

$$\frac{4}{8} - \frac{1}{8} = \frac{3}{8}$$

$$\frac{3}{8} - \frac{1}{8} = \frac{2}{8}$$

$$\frac{2}{8} - \frac{1}{8} = \frac{1}{8}$$

$$\frac{1}{8} - \frac{1}{8} = 0$$

Section 2 Common Fractions

Since 1/8 is subtracted 4 times from 1/2, 4 one-eighths are contained in 1/2.

$$\frac{1}{2} \div \frac{1}{8} = 4$$

The enlarged 1-inch scale shows four 1/8 inch parts in 1/2 inch.

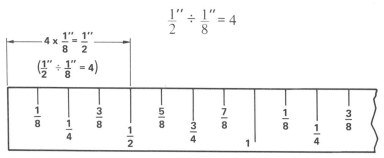

ENLARGED 1-INCH SCALE

Example: Divide $\frac{7''}{8}$ by $\frac{1''}{4}$.

Count the $\frac{1''}{4}$ in $\frac{7''}{8}$.

There are $3\frac{1}{2}$. *Ans*

The meaning of division of a fraction by a whole number, a fraction by a mixed number, and a mixed number by a mixed number is the same as division of fractions.

Practice

Substitute the values given for **a** and **b** and find how many times **a** is contained in **b**. Refer to 6-inch scale shown.

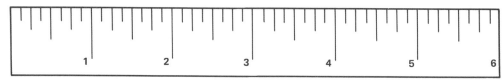

6-INCH SCALE

1. $a = 1\frac{1"}{2}, b = 3\frac{3"}{4}$ .... $2\frac{1}{2}$

2. $a = \frac{1"}{2}, b = 4\frac{3"}{4}$ .... $9\frac{1}{2}$

Exercise 11-1
Refer to the 6-inch scale shown. Substitute the values given for **a** and **b** in each of the problems 1–8 and find how many times **a** is contained in **b**.

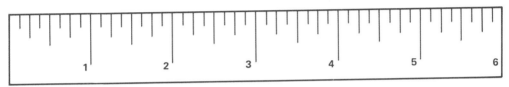

6-INCH SCALE

1. $a = \frac{1"}{4}, b = \frac{5"}{8}$ 5. $a = 1\frac{1"}{2}, b = 4\frac{1"}{2}$

2. $a = \frac{1"}{8}, b = \frac{5"}{8}$ 6. $a = \frac{3"}{4}, b = 1\frac{7"}{8}$

3. $a = \frac{1"}{4}, b = \frac{3"}{8}$ 7. $a = 2", b = 4\frac{1"}{2}$

4. $a = \frac{1"}{4}, b = \frac{7"}{8}$ 8. $a = 1\frac{1"}{4}, b = 3\frac{1"}{8}$

DIVISION OF FRACTIONS AS THE INVERSE OF MULTIPLICATION OF FRACTIONS

Division is the inverse of multiplication. Dividing by 2 is the same as multiplying by 1/2.

$$5 \div 2 = 2\frac{1}{2}$$

$$5 \times \frac{1}{2} = 2\frac{1}{2}$$

$$5 \div 2 = 5 \times \frac{1}{2}$$

Two is the inverse of 1/2 and 1/2 is the inverse of 2. The inverse of a fraction is a fraction which has the numerator and denominator interchanged. The inverse of 7/8 is 8/7.

DIVIDING FRACTIONS
To divide fractions invert the divisor, change to the inverse operation, and multiply.

Example: Divide. $\frac{5}{8} \div \frac{3}{4}$

Invert the divisor.
Multiply.

$$\frac{5}{8} \div \frac{3}{4} = \frac{5}{8} \times \frac{4}{3} = \frac{5}{6} \text{ Ans}$$

Practice

The machine bolt has a thread pitch of 1/16″. The pitch is the distance between 2 adjacent threads or the thickness of one thread. Find the number of threads in 7/8″.

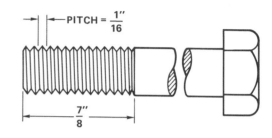

14 Ans

DIVIDING ANY COMBINATION OF FRACTIONS, MIXED NUMBERS, AND WHOLE NUMBERS

To divide any combination of fractions, mixed numbers, and whole numbers, write mixed numbers as fractions. Write whole numbers over the denominator 1. Invert the divisor. Change to the inverse operation and multiply.

Examples:

1. Divide. $10 \div \frac{4}{7}$

   Write the whole number 10 over the denominator 1.

   Invert the divisor, $\frac{4}{7}$.

   Change to the inverse operation and multiply.

   $$\frac{10}{1} \div \frac{4}{7} =$$

   $$\frac{10}{1} \times \frac{7}{4} = 17\frac{1}{2} \text{ Ans}$$

2. Divide. $\frac{19}{25} \div 3\frac{3}{10}$

   Write $3\frac{3}{10}$ as a fraction.

   Invert the divisor, $\frac{33}{10}$.

   Change to the inverse operation and multiply.

   $$\frac{19}{25} \div \frac{33}{10} =$$

   $$\frac{19}{25} \times \frac{10}{33} = \frac{38}{165} \text{ Ans}$$

3. Divide. $3\frac{3}{64} \div 40$

   Write $3\frac{3}{64}$ as a fraction. $\quad\quad\quad \frac{195}{64} \div \frac{40}{1} =$

   Write 40 over 1.
   Invert the divisor, $\frac{40}{1}$. $\quad\quad\quad \frac{195}{64} \times \frac{1}{40} = \frac{39}{512}$ *Ans*

   Change to the inverse
   operation and multiply.

4. Divide. $56\frac{2}{9} \div 8\frac{10}{21}$

   Write $56\frac{2}{9}$ and $8\frac{10}{21}$ as $\quad\quad\quad \frac{506}{9} \div \frac{178}{21} =$
   fractions.
   Invert the divisor, $\frac{178}{21}$. $\quad\quad\quad \frac{506}{9} \times \frac{21}{178} = 6\frac{169}{267}$ *Ans*

   Change to the inverse operation
   and multiply.

Practice
1. Shims each 7/16″ long are sheared (cut) from the length of strip stock shown. How many shims can be cut from this length?

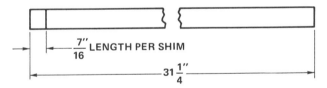

71 *Ans*

Note: The fractional part of a shim length is scrap and will not make another shim.

2. A baker prepares 412 1/2 pounds of bread dough. How many loaves can be cut from the dough if each loaf weighs 1 3/8 pounds? 300 *Ans*

Exercise 11-2
Divide the following fractions as indicated. Express the answers in lowest terms.

1. $\frac{1}{2} \div \frac{1}{4}$ $\quad\quad\quad$ 3. $\frac{2}{3} \div \frac{1}{3}$ $\quad\quad\quad$ 5. $\frac{1}{8} \div \frac{5}{8}$

2. $\frac{1}{4} \div \frac{1}{2}$ $\quad\quad\quad$ 4. $\frac{5}{8} \div \frac{1}{8}$ $\quad\quad\quad$ 6. $\frac{7}{16} \div \frac{7}{16}$

108 Section 2 Common Fractions

7. $\dfrac{4}{5} \div \dfrac{7}{8}$
8. $\dfrac{3}{16} \div \dfrac{3}{8}$
9. $\dfrac{9}{10} \div \dfrac{15}{27}$
10. $\dfrac{4}{5} \div \dfrac{9}{20}$
11. $\dfrac{8}{9} \div \dfrac{6}{7}$
12. $\dfrac{3}{7} \div \dfrac{9}{28}$

13. $\dfrac{1}{6} \div \dfrac{2}{5}$
14. $\dfrac{11}{12} \div \dfrac{5}{6}$
15. $\dfrac{9}{32} \div \dfrac{7}{8}$
16. $\dfrac{2}{3} \div \dfrac{9}{10}$
17. $\dfrac{18}{21} \div \dfrac{3}{8}$
18. $\dfrac{4}{9} \div \dfrac{1}{6}$

19. $\dfrac{11}{15} \div \dfrac{33}{40}$
20. $\dfrac{17}{20} \div \dfrac{6}{7}$
21. $\dfrac{8}{9} \div \dfrac{16}{17}$
22. $\dfrac{10}{11} \div \dfrac{4}{9}$
23. $\dfrac{15}{64} \div \dfrac{75}{128}$
24. $\dfrac{9}{35} \div \dfrac{19}{43}$

Exercise 11-3

Divide the following values as indicated. Express the answers in lowest terms.

1. $12 \div \dfrac{1}{2}$
2. $\dfrac{1}{2} \div 12$
3. $15 \div \dfrac{5}{8}$
4. $\dfrac{1}{9} \div 3$
5. $\dfrac{3}{4} \div 7$
6. $\dfrac{64}{65} \div 16$
7. $21 \div \dfrac{28}{31}$
8. $100 \div \dfrac{4}{5}$

9. $\dfrac{7}{16} \div 2\dfrac{1}{4}$
10. $2\dfrac{1}{4} \div \dfrac{7}{16}$
11. $3\dfrac{9}{32} \div \dfrac{11}{64}$
12. $5\dfrac{3}{5} \div \dfrac{15}{16}$
13. $\dfrac{19}{32} \div 10\dfrac{1}{4}$
14. $\dfrac{8}{15} \div 7\dfrac{27}{30}$
15. $16\dfrac{2}{3} \div \dfrac{2}{3}$
16. $\dfrac{1}{50} \div 9\dfrac{9}{10}$

17. $8\dfrac{5}{16} \div 2\dfrac{5}{8}$
18. $11\dfrac{4}{5} \div 10\dfrac{7}{10}$
19. $50 \div 15\dfrac{9}{25}$
20. $15\dfrac{9}{25} \div 50$
21. $7\dfrac{1}{2} \div 3\dfrac{5}{6}$
22. $25\dfrac{3}{4} \div 25\dfrac{1}{4}$
23. $6\dfrac{7}{8} \div 2\dfrac{2}{5}$
24. $2\dfrac{31}{32} \div 102\dfrac{3}{4}$

COMBINING ADDITION, SUBTRACTION, MULTIPLICATON, AND DIVISION OF FRACTIONS, MIXED NUMBERS, AND WHOLE NUMBERS

When solving a problem involving two or more different operations, first determine how the problem is solved. After the steps in the solution are thought through, follow the procedures for each operation.

Unit 11 Division of Common Fractions

Example:

A section of a wood frame building illustrating some of the structural members used in its construction is shown. Studs are the vertical timbers which form the frame for the walls and partitions of a wood structure. Joists are the horizontal members which support the floor and the ceiling for the level below. The veneer brick wall is made of bricks, 2 1/2" x 3 7/8" x 8 1/4", with 3/8" allowed for each mortar joint. The wall is 28 feet (336 inches) long and 10 1/32 feet (120 3/8 inches) high. The wall contains no windows or doors. Find the number of bricks needed for this wall. (No allowance is made for waste.)

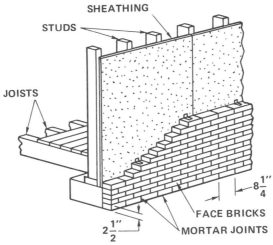

BRICK VENEER WALL SECTION

Solution: The number of bricks needed for the wall is equal to the number of bricks in one row times the number of rows.

Note: The number of mortar joints is one less than the number of bricks for both length and height.

Find the length of one brick and one mortar joint.

$$8\frac{1''}{4} + \frac{3''}{8} = 8\frac{5''}{8}$$

Subtract the one less mortar joint from the total length.

$$336'' - \frac{3''}{8} = 335\frac{5''}{8}$$

Find the number of bricks for one row.

$$335\frac{5''}{8} \div 8\frac{5''}{8} =$$

$$\frac{2\,685''}{8} \div \frac{69''}{8} =$$

$$\frac{2\,685''}{8} \times \frac{8}{69''} = 38\frac{63}{69} \text{ or 39 bricks}$$

Find the height of one brick and one mortar joint.	$2\frac{1''}{2} + \frac{3''}{8} = 2\frac{7''}{8}$	
Subtract the one less mortar joint from the total length.	$120\frac{3''}{8} - \frac{3''}{8} = 120''$	
Find the number of rows in the height.	$120'' \div 2\frac{7''}{8} =$	
	$120'' \div \frac{23''}{8} =$	
	$120'' \times \frac{8}{23''} = 41\frac{17}{23}$ or 42 rows	
Find the number of bricks needed for the wall.	$\frac{39 \text{ bricks}}{\text{row}} \times \frac{42 \text{ rows}}{1} = 1\,638$ bricks *Ans*	

UNIT REVIEW

Exercise 11-4

Divide the following fractions as indicated. Express the answers in lowest terms.

1. $\frac{1}{8} \div \frac{1}{4}$
2. $\frac{1}{4} \div \frac{1}{8}$
3. $\frac{3}{8} \div \frac{5}{8}$
4. $\frac{3}{5} \div \frac{3}{20}$

5. $\frac{6}{7} \div \frac{9}{35}$
6. $\frac{9}{32} \div \frac{5}{16}$
7. $\frac{19}{20} \div \frac{3}{5}$
8. $\frac{2}{15} \div \frac{11}{60}$

9. $\frac{5}{9} \div \frac{5}{6}$
10. $\frac{15}{64} \div \frac{9}{32}$
11. $\frac{5}{13} \div \frac{10}{11}$
12. $\frac{25}{128} \div \frac{5}{32}$

Divide the following values as indicated. Express the answers in lowest terms.

13. $9 \div \frac{1}{3}$
14. $\frac{1}{3} \div 9$
15. $\frac{7}{8} \div 6$
16. $150 \div \frac{15}{32}$

17. $3\frac{3}{4} \div \frac{7}{8}$
18. $2\frac{7}{32} \div \frac{9}{64}$
19. $\frac{7}{25} \div 1\frac{9}{10}$
20. $18\frac{1}{3} \div \frac{2}{3}$

21. $60 \div 14\frac{1}{4}$
22. $9\frac{1}{2} \div 3\frac{7}{8}$
23. $12\frac{3}{8} \div 12\frac{1}{8}$
24. $60\frac{5}{16} \div 10\frac{3}{32}$

Use the chart for problems 25–30. Find how much greater value **A** is than value **B**.

	A	B
25.	$\frac{7}{8}$ pounds ÷ $\frac{1}{8}$	$\frac{7}{8}$ pounds ÷ $\frac{3}{8}$
26.	16 inches ÷ $\frac{9}{32}$	10 inches ÷ $\frac{11}{16}$
27.	$320\frac{1}{2}$ gallons ÷ $\frac{1}{4}$	$512\frac{2}{3}$ gallons ÷ $\frac{3}{4}$
28.	$516\frac{3}{4}$ quarts ÷ $\frac{1}{2}$	$86\frac{3}{10}$ quarts ÷ $\frac{4}{5}$
29.	12 hours ÷ $2\frac{5}{6}$	24 hours ÷ $6\frac{1}{6}$
30.	$200\frac{7}{10}$ acres ÷ $20\frac{5}{8}$	$200\frac{7}{10}$ acres ÷ $20\frac{7}{8}$

PRACTICAL APPLICATIONS

Exercise 11-5

1. In order to drill the holes in the part shown a machinist finds the horizontal distances between the centers of the holes. There are five sets of holes, **A, B, C, D,** and **E**. The holes within each set are equally spaced.

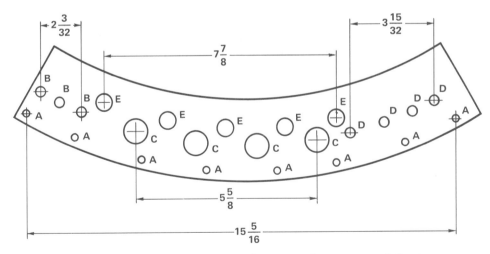

Find, in inches, the horizontal distance between the centers of the two consecutive holes listed.

a. **A** holes
b. **B** holes
c. **C** holes
d. **D** holes
e. **E** holes

2. A garment manufacturer purchases 1 500 yards of cotton goods. Each garment needs 3 7/8 yards. How many garments are produced from the 1 500 yards?

3. A plumber makes 4 pipe assemblies of different lengths. For any one assembly, the distances between two consecutive fittings, distance **B**, are equal as shown. Determine distance **B** for each pipe assembly.

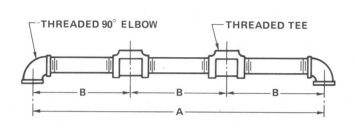

Pipe Assembly	A	B
#1	$21\frac{3}{8}''$	
#2	$28\frac{1}{2}''$	
#3	$32\frac{5}{16}''$	
#4	$37\frac{3}{4}''$	

4. A fast service restaurant chain sells the well known "quarter pounder." The quarter pounder is a hamburger containing 1/4 pound of ground beef. Five area restaurants use a total of 3 1/2 tons of ground beef in a certain week to make the quarter pounders. The restaurants are open 7 days a week. How many quarter pounders are sold per day by all five area restaurants?
Note: One ton equals 2 000 pounds.

5. A single-threaded square thread screw is shown. Square threads are used primarily for transmitting power. The lead of a screw is the distance that the screw advances in one turn (revolution). The lead is equal to the pitch in a single-threaded screw. The data listed in the table gives the number of turns required for the screw to advance a given number of inches. Find the leads for a–e.

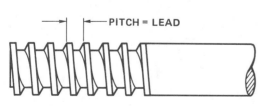

SINGLE-THREADED SQUARE THREAD SCREW

	SCREW ADVANCE	NUMBER OF TURNS	LEAD
a.	$2\frac{1}{4}''$	10	
b.	$7\frac{37}{64}''$	$24\frac{1}{4}$	
c.	$2\frac{7}{16}''$	$6\frac{1}{2}$	
d.	$1\frac{1}{2}''$	15	
e.	$6\frac{3}{10}''$	$12\frac{3}{5}$	

6. The point system is the standard of measure used by the printing industry. The point system is used for the measure of type, printing, and spacing materials. One point equals 1/72"; one pica equals 1/6".

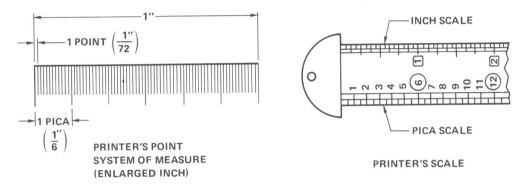

PRINTER'S POINT
SYSTEM OF MEASURE
(ENLARGED INCH)

PRINTER'S SCALE

Express each of the following inch measurements as points or picas.

a. 1/2" = ? picas
b. 1/8" = ? points
c. 3/8" = ? picas
d. 1/36" = ? points
e. 3 3/4" = ? picas
f. 1/4" = ? points

g. A sheet of paper 8 1/2" wide is divided in 20 equal width columns. What is the width of each column in picas?

7. A time-study analyzer collects the data on the production study chart shown.

	PRODUCTION STUDY CHART		
ITEM STUDIED: Part # ED-7612		Date: 5/23	
Employee	Daily Number of Pieces Produced	Number of Hours Worked	Pieces Produced Per Hour
A	310	$7\frac{3}{4}$	
B	375	$8\frac{1}{3}$	
C	424	$8\frac{5}{6}$	
D	315	$7\frac{1}{2}$	
		Average Hourly Production	

a. Find the number of pieces produced per hour by each employee.
b. Find the average hourly production of the 4 employees.

Note: The data on the production study sheet are valuable in determining production costs, unit pricing, and employee wage rates.

114   Section 2 Common Fractions

8. The gasoline consumption of an automobile is found in two categories, city driving and highway driving.

TYPE OF AUTOMOBILE	DISTANCE TRAVELED CITY	GASOLINE CONSUMED CITY	MILES PER GALLON CITY	DISTANCE TRAVELED HIGHWAY	GASOLINE CONSUMED HIGHWAY	MILES PER GALLON HIGHWAY
Full Size Sedan V-8	266 mi	$15\frac{1}{5}$ gal		406	$19\frac{1}{3}$ gal	
Intermediate Size – 6 Cyl.	262 mi	$13\frac{1}{10}$ gal		450	$18\frac{3}{4}$ gal	
Compact Size 4-Cyl.	300 mi	$11\frac{1}{4}$ gal		481	$14\frac{4}{5}$ gal	

a. Using the data listed in the table, find the gasoline consumption in miles per gallon (mi/gal) for each of the three types of automobiles.

b. How many more miles of highway driving can the compact 4-cylinder car travel than the V-8 engine full-size car when each car consumes 8 3/4 gallons of gasoline?

9. The structural column shown supports a portion of the ceiling of a building. The column is in compression or compressive stress; the load (portion of ceiling weight) tends to compress or crush the column. The load is distributed evenly over the complete cross-sectional area of the column. The unit compressive stress is the stress per square inch of cross-sectional area.

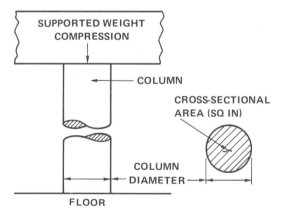

Unit stress is determined by dividing the load (weight being supported) by the cross-sectional area. Unit stress in this problem is found as the number of pounds per square inch. Calculations of unit stress are extremely important in mechanical and construction engineering and design. The table lists the data for 4 different size columns. Find the unit compressive stress for each column.

Unit 11 Division of Common Fractions 115

	CROSS-SECTIONAL AREA	LOAD (SUPPORTED WEIGHT)	UNIT COMPRESSIVE STRESS
a.	$3\frac{1}{7}$ square inches	$2\frac{1}{5}$ tons	
b.	$7\frac{3}{4}$ square inches	$6\frac{1}{5}$ tons	
c.	$28\frac{1}{2}$ square inches	$31\frac{7}{20}$ tons	
d.	$\frac{7}{8}$ square inch	$\frac{9}{10}$ ton	

Note: 1 ton = 2 000 pounds

10. An illustrator lays out a lettering job. Margins are measured and spacing for lettering is computed. Each row of lettering is 7/8" high with a 5/16" space between each row. Determine the number of rows of lettering on the sheet.

Hint: Notice that there is one less space between lettering rows than the number of rows.

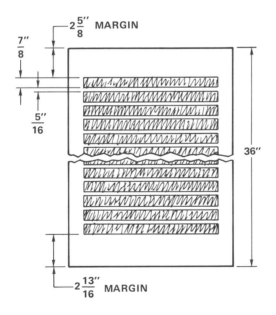

UNIT 12
Combined Operations with Common Fractions

OBJECTIVES

After studying this unit you should be able to

- Solve arithmetic expressions with combined operations by applying the proper order of operations.
- Solve practical combined operations problems involving formulas by applying the proper order of operations.

Combined operation problems given as arithmetic expressions require the use of the proper order of operations. Practical application problems based on formulas are found in occupational textbooks, handbooks, and manuals.

ORDER OF OPERATIONS

1. Do all the work in parentheses first. Parentheses are used to group numbers. In a problem expressed in fractional form, the numerator and the denominator are each considered as being enclosed in parentheses.

$$\frac{2\frac{5}{8} - \frac{3}{4}}{15 + 7\frac{9}{16}} = (2\frac{5}{8} - \frac{3}{4}) \div (15 + 7\frac{9}{16})$$

If an expression contains parentheses within brackets, do the work within the innermost parentheses first.

2. Do multiplication and division next in order from left to right.

3. Last, do addition and subtraction in order from left to right.

Unit 12 Combined Operations with Common Fractions 117

Examples:
1. Find dimension **Y** of the step block shown.

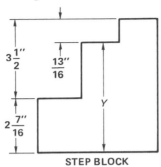

STEP BLOCK

Add. $2\frac{7''}{16} + 3\frac{1''}{2} = 5\frac{15''}{16}$

Subtract. $5\frac{15''}{16} - \frac{13''}{16} = 5\frac{1''}{8}$ *Ans*

2. Find the value of $\frac{2}{3} \times 6 \div 1\frac{1}{4}$

$$\frac{2}{3} \times 6 \div 1\frac{1}{4}$$

$$4 \div 1\frac{1}{4}$$

$$3\frac{1}{5} \; Ans$$

3. What is the total area (number of square feet) in the parcel of land shown?

Area = $120\frac{3'}{4} \times 80' + \frac{1}{2} \times 120\frac{3'}{4} \times 42\frac{1'}{3}$

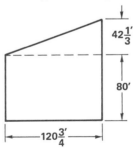

Area = $120\frac{3'}{4} \times 80' + \frac{1}{2} \times 120\frac{3'}{4} \times 42\frac{1'}{3}$

Area = 9 660 sq ft + $2\,555\frac{7}{8}$ sq ft

Area = $12\,215\frac{7}{8}$ sq ft *Ans*

Section 2 Common Fractions

4. Find the value of $\frac{3}{8} + 5\frac{1}{2} \times 8 - 12\frac{1}{4} \div 2$.

   $\frac{3}{8} + 44 - 12\frac{1}{4} \div 2$

   $\frac{3}{8} + 44 - 6\frac{1}{8}$

   $38\frac{1}{4}$ Ans

5. Find the value of $(\frac{3}{8} + 5\frac{1}{2}) \times 8 - 12\frac{1}{4} \div 2$.

   $5\frac{7}{8} \times 8 - 12\frac{1}{4} \div 2$

   $47 - 6\frac{1}{8}$

   $40\frac{7}{8}$ Ans

6. Find the value of $\dfrac{84\frac{3}{5} - 18\frac{7}{10} \times 3}{1\frac{8}{9} + 20\frac{2}{3} \div 2\frac{2}{5}} - 1\frac{4}{7}$.

   $(84\frac{3}{5} - 18\frac{7}{10} \times 3) \div (1\frac{8}{9} + 20\frac{2}{3} \div 2\frac{2}{5}) - 1\frac{4}{7}$

   $28\frac{1}{2} \div 10\frac{1}{2} - 1\frac{4}{7}$

   $2\frac{5}{7} - 1\frac{4}{7} = 1\frac{1}{7}$ Ans

Practice

1. A semicircular-sided section is fabricated in a sheet metal shop. A semicircle is a half circle which is formed on sheet metal parts by rolling. The sheet metal worker makes a stretchout of the section. A stretchout is a flat layout which when formed makes the required part.

   Length Size = $3\frac{1}{7} \times d + 2 \times w$ where d = diameter of roll (semicircle)
   w = distance between centers of semicircles

   Length of each roll = $\dfrac{3\frac{1}{7} \times d}{2}$

   Location of roll = $\dfrac{w}{2}$

Unit 12 Combined Operations with Common Fractions 119

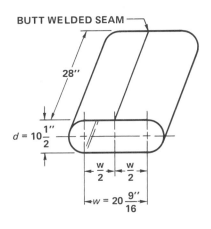
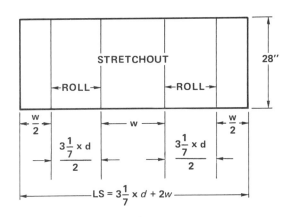

a. Find, in inches, the length of the stretchout. $74\frac{1}{8}''$ Ans

b. Find, in inches, the length of each roll. $16\frac{1}{2}''$ Ans

c. Find the location of each roll. $10\frac{9}{32}''$ Ans

2. In the aircraft industry, the weight of a part is an important factor in mechanical design. The wedge-shaped part shown is a component of a design assembly. It is machined from a solid block of magnesium alloy which weighs 1/16 pound per cubic inch. Find the weight of the part.

$$\text{Volume of wedge} = \frac{(2 \times a + c) \times b \times h}{6}$$

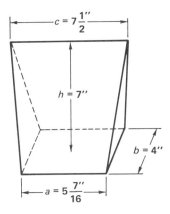

$5\frac{23}{64}$ pounds Ans

Section 2 Common Fractions

UNIT REVIEW
Exercise 12-1
Perform the indicated operations.

1. $\frac{3}{4} + \frac{1}{2} - \frac{1}{8}$

2. $3 - \frac{7}{8} + 2\frac{3}{8}$

3. $\frac{7}{10} + \frac{3}{5} - \frac{2}{15}$

4. $2\frac{2}{3} + 12 - \frac{5}{6} + \frac{7}{12}$

5. $(\frac{5}{8} + \frac{3}{4}) \div \frac{1}{2}$

6. $\frac{5}{8} + \frac{3}{4} \div \frac{1}{2}$

7. $\frac{16}{\frac{3}{4}} - 2\frac{7}{16}$

8. $\dfrac{16 - 2\frac{7}{16}}{\frac{3}{4}}$

9. $\dfrac{\frac{3}{5} + \frac{9}{10} - \frac{4}{25}}{4}$

10. $\dfrac{5\frac{5}{8}}{2\frac{1}{4} - 1\frac{1}{8}}$

11. $25\frac{2}{3} - 18\frac{5}{6} + \dfrac{6\frac{1}{2}}{1\frac{1}{2}}$

12. $10\frac{31}{64} + \frac{9}{16} \times 4 - 1\frac{1}{2}$

13. $10\frac{31}{64} + \frac{9}{16} \times (4 - 1\frac{1}{2})$

14. $(\frac{1}{25} \times 10) \div (\frac{3}{10} \times 20)$

15. $(\frac{1}{25} \times 10) \div \frac{3}{10} + 20$

16. $\dfrac{7\frac{3}{4} + 3 \times \frac{1}{2}}{10 - 5\frac{1}{4}}$

17. $\dfrac{(7\frac{3}{4} + 3) \times \frac{1}{2}}{10 - 5\frac{1}{4}}$

18. $(\frac{3}{16} - \frac{3}{32}) \times \frac{1}{8} \div \frac{1}{16}$

19. $(10\frac{2}{3} + 5\frac{1}{3} \times \frac{5}{6}) \div 8\frac{1}{6}$

20. $\frac{7}{15} \times (4 - \frac{7}{10}) + 5 \times 12\frac{2}{5}$

21. $(\frac{7}{15} \times 4 - \frac{7}{10} + 5) \times 12\frac{2}{5}$

22. $\dfrac{9\frac{1}{2} \times (3\frac{7}{8} - 2\frac{3}{8})}{3\frac{1}{4}}$

23. $\dfrac{30}{3\frac{1}{3}} - \dfrac{24\frac{3}{8}}{3}$

24. $\dfrac{120 - 20\frac{1}{2} \times 4}{40\frac{2}{3} + 12 - 6\frac{1}{2}}$

25. $\dfrac{(120 - 20\frac{1}{2}) \times 4}{40\frac{2}{3} + 12 - 6\frac{1}{2}}$

26. $8\frac{3}{4} \times 10 + 50\frac{7}{10} \div (12 + \frac{2}{3})$

PRACTICAL APPLICATIONS

Exercise 12-2

1. The horsepower (hp) of a motor is found using the following formula:

$$hp = \frac{2 \times 3\frac{1}{7} \times T \times r/min}{33\,000}$$ where T = torque

Torque is a turning effect. It is the tendency of the armature to rotate. Torque is expressed in pound-feet (lb-ft). Using the torque and revolutions per minute of 4 electric motors, find the horsepower of each.

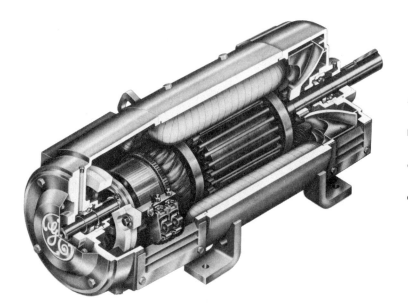

	TORQUE (lb-ft)	r/min	hp
a.	$3\frac{1}{2}$	2 250	
b.	$4\frac{3}{8}$	2 400	
c.	$5\frac{1}{4}$	2 750	
d.	$4\frac{2}{3}$	3 750	

Electric Motor (General Electric Company)

2. A simple parallel electrical circuit is shown. An environmental systems technician finds the total resistance of the circuit.

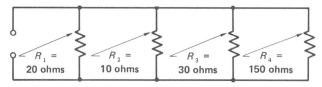

$R_1 = $ 20 ohms $R_2 = $ 10 ohms $R_3 = $ 30 ohms $R_4 = $ 150 ohms

$$R_T = \frac{1}{\frac{1}{R_1} + \frac{1}{R_2} + \frac{1}{R_3} + \frac{1}{R_3}}$$

where R_1, R_2, R_3, R_4 are the individual resistors and R_T is the total resistance.

Find the total resistance of the circuit.

Note: Compare your answer for R_T to the values of the individual resistors. Observe that the total resistance is less than the smallest resistor (R_2 = 10 ohms). In a parallel circuit, adding resistors reduces total resistance.

3. A sheet metal section is shown. Find, in inches, the length of the stretchout.

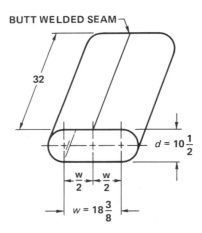

4. In the manufacturing industry, many pieces of the same part are produced. In estimating the cost of production per piece, one factor considered is cutting time. Cutting time for a drilling operation is determined by the depth to be drilled, the revolutions per minute of the drill, and the tool feed. The tool feed is the depth of material cut for each revolution of the drill. A manufacturing engineering technician in estimating the drilling time per piece of a production run uses the following formula:

$T = \dfrac{L}{F \times N}$ where T = cutting time in minutes per drilled hole.
L = length (depth) of hole in inches.
F = tool (drill) feed in inches per revolution.
N = revolutions per minute (r/min) of the drill.

Hole	F	L	N	T
#1	$\frac{1''}{32}$	$3\frac{3}{4}''$	300	
#2	$\frac{1''}{64}$	$2\frac{1}{2}''$	320	
#3	$\frac{1''}{40}$	$4\frac{3}{4}''$	380	
#4	$\frac{1''}{50}$	$1\frac{3}{4}''$	350	
#5	$\frac{1''}{40}$	$3\frac{1}{2}''$	400	

a. Determine the cutting time for each of the 5 drilled holes in a piece.
b. Determine the total cutting time per piece for all 5 drilled holes.

Unit 12 Combined Operations with Common Fractions 123

5. Pulleys and belts are widely used in automotive, commerical, and industrial equipment. An application of a belt drive in an automobile is shown.

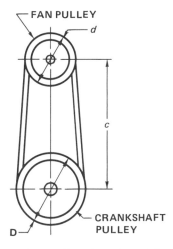

Determine the belt lengths required for each of the 4 belt drives listed in the table.

$$L = \frac{D + d}{2} \times 3\frac{1}{7} + 2 \times c$$

where L = length of belt in inches
D = diameter of large pulley in inches
d = diameter of small pulley in inches
c = center-to-center distance of pulleys in inches

	D	d	c	L
a.	$5\frac{1''}{2}$	$3\frac{1''}{4}$	$10\frac{1''}{4}$	
b.	$6\frac{3''}{8}$	$4\frac{1''}{8}$	$11\frac{1''}{16}$	
c.	$4\frac{1''}{16}$	$2\frac{15''}{16}$	$8\frac{5''}{8}$	
d.	$6\frac{13''}{16}$	$5\frac{7''}{16}$	$9\frac{1''}{8}$	

6. Hourly paid employees in many companies get time-and-a-half pay for overtime hours worked during their regular workweek. Overtime pay is based on the number of hours worked over the normal workweek hours. The normal workweek varies from company to company, but is usually from 36 to 40 hours. Time-and-a-half means the employee's overtime rate of pay is 1 1/2 times the normal rate of pay or the employee is credited with 1 1/2 hours for each hour worked. Find the total number of hours credited to each employee.

Total number of hours credited = Normal workweek hours + (Number of hours worked − Normal workweek hours) × $1\frac{1}{2}$.

	Employee			
	A	B	C	D
Normal workweek hours	38	40	36	40
Hours worked during week	$44\frac{1}{2}$	$47\frac{1}{4}$	$39\frac{3}{4}$	$46\frac{1}{2}$
Total hours credited				

7. A tank is to be fabricated from steel plate. The specifications call for 3-gauge (approx. 1/4" thick) plate.

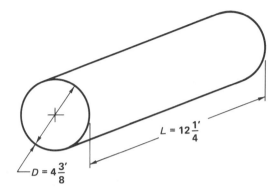

An engineering aide refers to a metals handbook and finds one square foot of 3-gauge steel plate weighs 9 1/2 pounds per square foot. The weight of the tank is computed using the following formula:

Weight of tank in pounds = $3\frac{1}{7} \times D \times (\frac{1}{2} \times D + L) \times W$

where D = diameter of tank in feet
       L = length of tank in feet
       W = weight of 1 square foot of plate

Find the weight of the tank.

8. A landscaper contracts to grade and plant ground cover in the shaded portion of the plot of land shown. One step in determining the total cost of the job is to find the area to be graded and planted. The following formula is used to find the area:

$$\text{Area} = R \times R - \frac{11 \times R \times R}{14}$$

Determine the area in square feet of ground to be graded and planted.

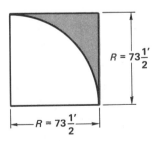

SECTION 3
Decimal Fractions

UNIT 13
Introduction to Decimal Fractions

OBJECTIVES

After studying this unit you should be able to

- Express decimal values of distances shown on a graduated line segment.
- Write decimal numbers in word form.
- Write numbers expressed in word form as decimal fractions.
- Express common fractions having denominators of powers of ten as equivalent decimal fractions.

Calculations using decimals are often faster and easier to make than fractional computations. The decimal system of measurement is widely used in occupations where greater precision than fractional parts of an inch is required. Decimals are used to compute to any required degree of precision. Certain industries require a degree of precision to the millionths of an inch.

Most machined parts are manufactured using decimal system dimensions and decimal machine settings. The electrical and electronic industries generally compute and measure using decimals. Computations required for the design of buildings, automobiles, and aircraft are based on the decimal system. Occupations in the retail, wholesale, office, health, transportation, and communications fields require decimal calculations. Finance and insurance companies base their computational procedures on the decimal system. Our monetary system of dollars and cents is based on the decimal system.

Electronic hand calculators, such as the one shown are designed for entry of decimal numbers. Answers are given in decimal form. Business and industry use a variety of different types of electronic calculators.

Handheld Calculator (Monroe, The Calculator Company)

EXPLANATION OF DECIMAL FRACTIONS

A decimal fraction is not written as a common fraction with a numerator and denominator. The decimal fraction is written using a decimal point. Decimal fractions are equivalent to common fractions having denominators which are powers of ten. *Powers of 10* are numbers which are obtained by multiplying 10 by itself a certain number of times. Numbers such as 100; 1 000; 10 000; 100 000; and 1 000 000 are powers of ten.

MEANING OF FRACTIONAL PARTS

The line segment shown is 1 unit long. It is divided in 10 equal smaller parts. The locations of common fractions and their decimal fraction equivalents are shown on the line.

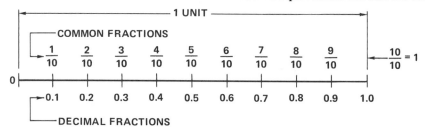

1 UNIT LINE

One of the ten equal small parts, 1/10 or 0.1 of the 1 unit line is greatly enlarged. The 1/10 or 0.1 unit is divided into 10 smaller units. The locations of common fractions and their decimal fraction equivalents are shown on this line.

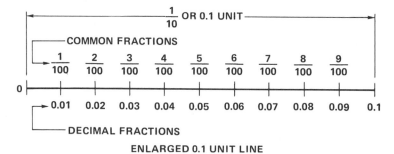

ENLARGED 0.1 UNIT LINE

If the 1/100 or 0.01 division is divided into 10 equal smaller parts, the resulting parts are 0.001, 0.002, 0.003, 0.004, 0.005, 0.006, 0.007, 0.008, 0.009, and 0.01. Each time a decimal point is moved one place to the left, a value 1/10 or 0.1 of the previous value is obtained. Each time a decimal point is moved one place to the right, a value 10 times greater than the previous value is obtained.

Exercise 13-1

1. Write the decimal value of distances A–E.

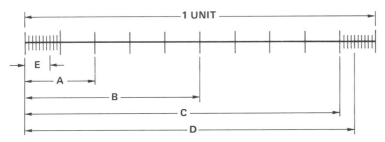

2. Write the decimal value of distances F–J.

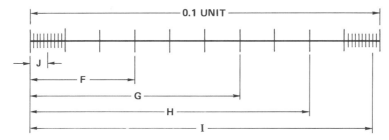

READING DECIMAL FRACTIONS

The chart shown gives the names of the parts of a number with respect to their positions from the decimal point.

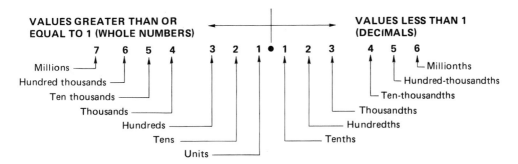

To read a decimal, read the number as a whole number. Then say the name of the decimal place of the last digit to the right.

Section 3 Decimal Fractions

Examples:

1. 0.43 is read, "forty-three hundredths."
2. 0.532 is read, "five hundred thirty-two thousandths."
3. 0.002 8 is read, "twenty-eight ten-thousandths."
4. 0.280 0 is read, "two thousand eight hundred ten-thousandths."

To read a mixed decimal (a whole number and a decimal fraction), read the whole number, read the word "and" at the decimal point, and read the decimal.

Examples:

1. 2.65 is read, "two and sixty-five hundredths."
2. 9.002 is read, "nine and two thousandths."
3. 135.078 7 is read, "one hundred thirty-five and seven hundred eighty-seven ten-thousandths."

Exercise 13-2

Write the following numbers as words.

1. 0.3	6. 0.018	11. 0.000 5	16. 351.032
2. 0.03	7. 0.07	12. 0.002 09	17. 33.333 3
3. 0.30	8. 0.009 8	13. 15.876	18. 299.000 9
4. 0.300	9. 0.105	14. 1.084	19. 299.090 0
5. 0.175	10. 0.015	15. 27.002 7	20. 158.800 8

Write the following words as decimals or mixed decimals.

21. nine tenths
22. six hundredths
23. eighty-one hundredths
24. two ten-thousandths
25. four hundred thirty-five thousandths
26. sixty-six ten-thousandths
27. six hundred ten-thousandths
28. thirty-one hundred-thousandths
29. three hundred one hundred thousandths
30. nine and twelve hundredths
31. twelve and one thousandth
32. twelve thousandths
33. twenty and twenty ten-thousandths
34. six and thirty-five ten-thousandths
35. two and sixty-two hundredths
36. sixty-two and two hundredths

SIMPLIFIED METHOD OF READING DECIMAL FRACTIONS

Often a simplified method of reading decimal fractions is used in actual on-the-job applications. This method is generally quicker, easier, and less likely to be misinterpreted. A tool and die maker reads 0.018 7 inches as "point zero, one, eight, seven inches." An electronics technician reads 2.125 amperes as "two, point one, two, five amperes."

WRITING DECIMAL FRACTIONS

A common fraction with a denominator which is a power of ten can be written as a decimal fraction. Replace the denominator with a decimal point. The decimal point is placed to the left of the first digit of the numerator. There are as many decimal places as there are zeros in the denominator. When writing a decimal fraction it is advisable to place a zero to the left of the decimal point.

Examples:
Write each common fraction as a decimal fraction.

1. $\frac{7}{10}$ = 0.7 *Ans*   There is 1 zero in 10 and 1 decimal place in 0.7.

2. $\frac{65}{100}$ = 0.65 *Ans*   There are 2 zeros in 100 and 2 decimal places in 0.65.

3. $\frac{793}{1\,000}$ = 0.793 *Ans*   There are 3 zeros in 1 000 and 3 decimal places in 0.793.

4. $\frac{9}{10\,000}$ = 0.000 9 *Ans*   There are 4 zeros in 10 000 and 4 decimal places in 0.000 9. In order to maintain proper place value, 3 zeros are written between the decimal point and the 9.

Exercise 13-3

Each of the following common fractions has a denominator which is a power of 10. Write the equivalent decimal fraction for each.

1. $\frac{9}{10}$
2. $\frac{19}{100}$
3. $\frac{197}{1\,000}$
4. $\frac{3}{100}$
5. $\frac{3}{1\,000}$
6. $\frac{27}{1\,000}$
7. $\frac{83}{100}$
8. $\frac{1}{100}$
9. $\frac{323}{1\,000}$
10. $\frac{287}{10\,000}$
11. $\frac{41}{1\,000}$
12. $\frac{77}{10\,000}$
13. $\frac{999}{1\,000}$
14. $\frac{1}{1\,000}$
15. $\frac{1}{10\,000}$
16. $\frac{8\,111}{10\,000}$
17. $\frac{7\,663}{100\,000}$
18. $\frac{3}{100\,000}$
19. $\frac{424\,871}{1\,000\,000}$
20. $\frac{1}{1\,000\,000}$

130 Section 3 Decimal Fractions

UNIT REVIEW

Exercise 13-4

Write the equivalent decimal fraction for each of the following common fractions.

1. $\frac{3}{10}$
2. $\frac{87}{100}$
3. $\frac{1}{100}$
4. $\frac{219}{1\,000}$
5. $\frac{227}{10\,000}$
6. $\frac{2\,227}{10\,000}$
7. $\frac{511}{100\,000}$
8. $\frac{3}{100\,000}$
9. $\frac{9\,999}{10\,000}$
10. $\frac{5\,281}{100\,000}$
11. $\frac{7}{10\,000}$
12. $\frac{77}{10\,000}$
13. $\frac{777}{10\,000}$
14. $\frac{863\,333}{1\,000\,000}$
15. $\frac{19}{1\,000\,000}$
16. $\frac{1\,999}{1\,000\,000}$

17. Write the decimal value of distances A–E.

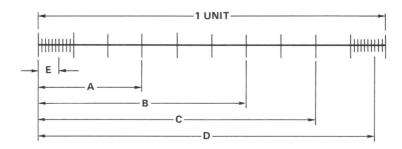

18. Write the decimal value of distances F–J.

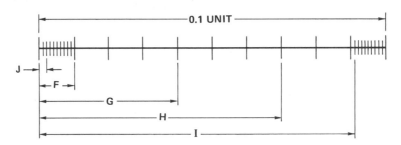

Write the following numbers as words.

19. 0.9
20. 0.06
21. 0.85
22. 0.298
23. 0.013
24. 0.003
25. 0.008 8
26. 0.712
27. 0.000 4
28. 0.003 22
29. 2.923
30. 13.013
31. 595.866
32. 77.000 4
33. 304.020 6
34. 1 815.356 32

Write the following words as decimals or mixed decimals.

35. four tenths
36. eight hundredths
37. seventy-six hundredths
38. one hundred thousandths
39. three ten-thousandths
40. three hundred ten-thousandths
41. seven and nine tenths
42. nineteen and nineteen hundredths
43. eleven and eight thousandths
44. twenty-two and fifty thousandths
45. five hundred and five hundredths
46. one and forty-one ten-thousandths

Write the following decimals and mixed decimals as words using the simplified method.

47. 0.293
48. 0.03
49. 0.706
50. 0.100 2
51. 0.700 79
52. 5.120 3
53. 3.040
54. 8.010 05

UNIT 14
Equivalent Decimal and Common Fractions

OBJECTIVES

After studying this unit you should be able to

- Round decimal fractions to any indicated number of places.
- Express common fractions as decimal fractions.
- Express decimal fractions as common fractions.

When working with decimals, the computations and answers may contain more decimal places than are needed. The number of decimal places needed depends on the degree of precision desired. The degree of precision depends on how the obtained decimal value is going to be used. The tools, machinery, equipment, and materials determine the degree of precision which can be obtained.

It is not realistic for a carpenter to attempt to saw a board to a 6.251 8-inch length. The 6.251 8-inch-length is realistic in the machine trades. A surface grinder operator can grind a metal part to four decimal place precision.

ROUNDING DECIMAL FRACTIONS

To round a decimal fraction, locate the digit in the number that gives the desired degree of precision. Increase that digit by 1 if the digit which directly follows is 5 or more. Do not change the value of the digit if the digit which follows is less than 5. Drop all digits which follow.

Examples:

1. A designer computes a dimension of 0.738 62 inch. Three place precision is needed for the part which is being drawn.

   Locate the digit in the third decimal place. (8)

   The fourth decimal place digit, 6, is greater than 5 and increases 8 to 9.

   0.738 62 inch ≈ 0.739 inch *Ans*

2. In determining rivet hole locations, a sheet-metal technician computes a dimension of 1.503 8 inches. Two place precision is needed for laying out the hole locations.

   Locate the digit in the second decimal place. (0)

   The third decimal place digit, 3, is less than 5 and does not change the value, 0.

   1.503 8 inches ≈ 1.50 inches *Ans*

Practice

Round each of the following numbers to the indicated number of decimal places.

1. 0.386 (2 places) 0.39
2. 1.893 (1 place) 1.9
3. 0.533 32 (3 places) 0.533
4. 29.323 9 (2 places) 29.32

Exercise 14-1

Round each of the following numbers to the indicated number of decimal places.

1. 0.837 (2 places)
2. 0.344 (2 places)
3. 0.873 (1 place)
4. 0.007 2 (3 places)
5. 0.007 2 (2 places)
6. 0.007 2 (1 place)
7. 0.449 9 (2 places)
8. 0.888 8 (3 places)
9. 0.051 5 (1 place)
10. 0.014 97 (4 places)
11. 22.195 5 (3 places)
12. 108.305 6 (2 places)
13. 831.400 19 (4 places)
14. 89.899 4 (3 places)
15. 89.899 5 (3 places)
16. 511.077 (1 place)
17. 13.005 1 (2 places)
18. 722.010 10 (3 places)
19. 48.358 78 (4 places)
20. 100.999 9 (1 place)

EXPRESSING COMMON FRACTIONS AS DECIMAL FRACTIONS

Expressing common fractions as decimal fractions is used in many occupations. A bookkeeper in working a financial statement expresses $16 3/20 as $16.15. In preparing a medication, a nurse may express 1 1/5 ounces of solution as 1.2 ounces.

A common fraction is an indicated division. A common fraction is expressed as a decimal fraction by dividing the numerator by the denominator.

Example: Express $\frac{3}{8}$ as a decimal fraction.

Write $\frac{3}{8}$ as an indicated division.

$$\begin{array}{r} 0.375 \; Ans \\ 8 \overline{) 3.000} \end{array}$$

Place a decimal point after the 3 and add zeros to the right of the decimal point. Note: Adding zeros after the decimal point does not change the value of the dividend; 3 has the same value as 3.000.

Place the decimal point for the answer directly above the decimal point in the dividend. Divide.

A common fraction which divides evenly is expressed as an even or *terminating decimal*. A common fraction which will not divide evenly is expressed as a repeating or *nonterminating decimal*.

Example: Express $\frac{2}{3}$ as a decimal.

Write $\frac{2}{3}$ as an indicated division.

$$\begin{array}{r} 0.666\;6\ldots \; Ans \\ 3 \overline{) 2.000\;0} \end{array}$$

Place a decimal point after the 2 and add zeros to the right of the decimal point.

Place the decimal point for the answer directly above the decimal point in the dividend. Divide.

Exercise 14-2

Express each of the following common fractions as decimal fractions. Where necessary, round the answers to 4 decimal places.

1. $\dfrac{1}{2}$
2. $\dfrac{1}{4}$
3. $\dfrac{5}{8}$
4. $\dfrac{7}{8}$
5. $\dfrac{13}{32}$

6. $\dfrac{1}{3}$
7. $\dfrac{4}{5}$
8. $\dfrac{10}{11}$
9. $\dfrac{2}{3}$
10. $\dfrac{5}{6}$

11. $\dfrac{1}{25}$
12. $\dfrac{3}{25}$
13. $\dfrac{47}{64}$
14. $\dfrac{19}{32}$
15. $\dfrac{1}{16}$

16. $\dfrac{7}{32}$
17. $\dfrac{20}{21}$
18. $\dfrac{19}{20}$
19. $\dfrac{99}{101}$
20. $\dfrac{23}{24}$

21. Determine dimensions A, B, C, and D as decimal fractions to 4 decimal places.

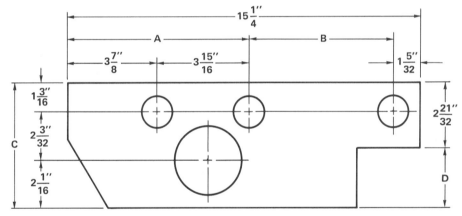

SUPPORT BRACKET

DECIMAL FRACTIONAL PARTS PROBLEMS

To find decimal answers for certain problems, it is necessary to find the answers using two or more steps. The answers are first computed as common fractions; then the common fractions are expressed as decimal fractions.

> Example: A forester estimates that 300 acres of a 2 500-acre forest are destroyed by fire. What decimal fraction of the forest is destroyed?
>
> Write the common fraction which compares the number of acres destroyed with the total number of acres.
>
> $\dfrac{300 \text{ acres}}{2\,500 \text{ acres}} = 0.12$ *Ans*
>
> Express the common fraction as a decimal fraction.

136 Section 3 Decimal Fractions

Practice

In this circuit, the total current is the sum of the currents ($I_1 + I_2 + I_3 + I_4$). What decimal fraction of the total current (amperes) in the circuit shown is received by resistance #2 (R_2)? Round the answer to 3 decimal places?

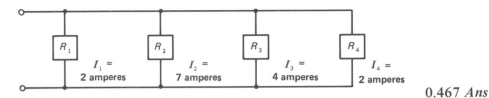

0.467 *Ans*

Exercise 14-3

Determine the decimal fraction answers for each of the following problems. Where necessary, round the answers to 3 decimal places.

1. A building contractor determines the total cost of a job as $54 500. Labor costs are $21 800. What decimal fraction of the total cost is the labor cost?

2. The displacement of an automobile engine is 246 cubic inches. The engine is rebored an additional 4 cubic inches. What decimal fraction of the displacement of the rebored engine is the displacement of the engine before rebore?

3. A mason lays a sidewalk to the dimensions shown.

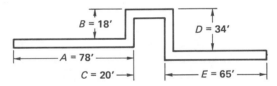

   a. What decimal fraction of the total length of sidewalk is distance A?
   b. What decimal fraction of the total length of sidewalk is distance C?
   c. What decimal fraction of the total length of sidewalk is distance E?
   d. What decimal fraction of the total length of sidewalk is distance B plus distance D?

4. The interior walls of a house contain a total area of 3 050 square feet. A painter and decorator paints 1 800 square feet. What decimal fraction of the total wall area is the area whch remains to be painted?

5. A hospital dietitian allows 700 calories for a patient's breakfast and 750 calories for lunch. The total daily intake is 2 500 calories.

   a. What decimal fraction of the total daily calorie intake is allowed for breakfast?
   b. What decimal fraction of the total daily calorie intake is allowed for lunch?
   c. What decimal fraction of the total daily calorie intake is allowed for dinner?

6. In pricing merchandise, retail firms sometimes use the following simple formula.

   Retail price = cost of goods + overhead expenses + desired profit

   Retail price is the price the customer is charged. Cost of goods is the price the retailer pays the manufacturer or supplier.

ITEM	RETAIL PRICE	COST OF GOODS	OVERHEAD EXPENSES	DESIRED PROFIT
A		$325	$105	$28
B		$120	$ 36	$ 8
C	$930	$672	$212	

a. What decimal fraction of the retail price is the desired profit of Item **A**?
b. What decimal fraction of the retail price is the desired profit of Item **B**?
c. What decimal fraction of the retail price is the desired profit of Item **C**?
d. What decimal fraction of the retail price is the cost of goods of Item **A**?
e. What decimal fraction of the retail price is the overhead expenses of Item **B**?

EXPRESSING DECIMAL FRACTIONS AS COMMON FRACTIONS

Dimensions given or computed as decimals are often expressed as common fractions for on-the-job measurements. A carpenter expresses 10.625" as 10 5/8" when measuring the length to saw a board. In locating bolt holes on a beam, a structural ironworker changes a dimension given as 12'-6.75" to 12'-6 3/4".

To change a decimal fraction to a common fraction, write the number after the decimal point as the numerator of a common fraction. Write the denominator as 1 followed by as many zeros as there are digits to the right of the decimal point. Express the common fraction in lowest terms.

Examples:

1. Express 0.7 as a common fraction.

   Write 7 as the numerator.

   Write the denominator as 1 followed by 1 zero. The denominator is 10.

   $\frac{7}{10}$ Ans

2. Express 0.065 as a common fraction.

   Write 65 as the numerator.

   Write the denominator as 1 followed by 3 zeros. The denominator is 1 000.

   Express the fraction in lowest terms.

   $\frac{65}{1\,000} = \frac{13}{200}$ Ans

Exercise 14-4

Express each of the following decimal fractions as common fractions. Express the answers in lowest terms.

1. 0.3
2. 0.75
3. 0.33
4. 0.42
5. 0.325
6. 0.050
7. 0.005
8. 0.801
9. 0.903
10. 0.36
11. 0.028
12. 0.654
13. 0.010 8
14. 0.007 6
15. 0.999
16. 0.000 8
17. 0.937 5
18. 0.009 99
19. 0.437 5
20. 0.031 25

21. Express, in inches, dimensions A–F as common fractions.

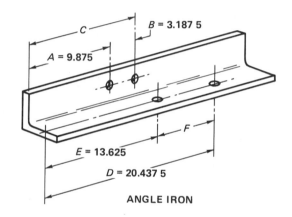

ANGLE IRON

UNIT REVIEW

Exercise 14-5

Round each of the following numbers to the indicated number of decimal places.

1. 0.943 (2 places)
2. 0.175 (2 places)
3. 0.009 6 (3 places)
4. 0.649 (1 place)
5. 0.650 (1 place)
6. 0.007 3 (1 place)
7. 0.777 7 (3 places)
8. 0.053 (1 place)
9. 17.043 (1 place)
10. 34.135 5 (3 places)
11. 99.905 1 (2 places)
12. 306.300 06 (4 places)
13. 99.999 (2 places)
14. 705.020 25 (3 places)
15. 51.329 96 (4 places)
16. 14.008 5 (2 places)

Unit 14 Equivalent Decimal and Common Fractions 139

Express each of the following common fractions as decimal fractions. Where necessary, round the answers to 3 decimal places.

17. $\dfrac{3}{8}$ 21. $\dfrac{5}{9}$ 25. $\dfrac{17}{32}$ 29. $\dfrac{9}{10}$

18. $\dfrac{2}{3}$ 22. $\dfrac{9}{11}$ 26. $\dfrac{33}{64}$ 30. $\dfrac{100}{103}$

19. $\dfrac{7}{8}$ 23. $\dfrac{1}{6}$ 27. $\dfrac{15}{16}$ 31. $\dfrac{8}{15}$

20. $\dfrac{1}{32}$ 24. $\dfrac{49}{50}$ 28. $\dfrac{3}{7}$ 32. $\dfrac{41}{48}$

33. A drill jig is shown. Find, in inches, dimensions A, B, C, D, and E. Express the answers as decimal fractions to 4 decimal places.

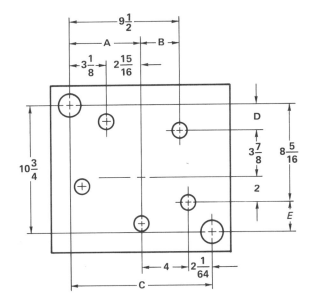

Express the following decimal fractions as common fractions. Express the answer in lowest terms.

34. 0.6   38. 0.062 5   42. 0.156 25   46. 0.090
35. 0.250   39. 0.058   43. 0.375   47. 0.998
36. 0.860   40. 0.002   44. 0.001 5   48. 0.218 75
37. 0.187 5   41. 0.908   45. 0.001 9   49. 0.000 86

PRACTICAL APPLICATIONS

Exercise 14-6

Determine the decimal fraction answer for each of the following problems. Where necessary, round the answers to 2 decimal places.

1. A clothing manufacturer purchases 75 000 yards of cloth for a certain order. After using 23 000 yards, the order is cancelled. What decimal fraction of the number of yards purchased is the number of yards actually used?

2. Five pieces are cut from the length of round stock shown. After the pieces are cut, the remaining length is thrown away. What decimal fraction of the original length of round stock (17") is the length which is thrown away? All dimensions are in inches.

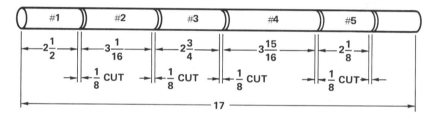

3. To complete a job of printing circulars, a printer needs a total of 11 hours. Composition takes 6 1/4 hours and lock-up and make-ready takes 3/4 hour. The rest of the time is taken in feeding the job. What decimal fraction of the total job time is needed for feeding?

4. A carpenter estimates the number of lengths of 2 x 4's needed for a building. The table lists the estimated requirements and the number of each length in stock.

	10' LENGTHS	12' LENGTHS	16' LENGTHS
Estimated number required for job	110	150	90
Number in stock	58	65	42

    a. What decimal fraction of the number of 2 x 4's of each length required for the job is the number of each length in stock?

    b. What decimal fraction of the total of all 2 x 4's required for the job is the number of all 2 x 4's which must be ordered?

UNIT 15
Addition and Subtraction of Decimal Fractions

OBJECTIVES

After studying this unit you should be able to

- Add decimal fractions.
- Subtract decimal fractions.
- Solve practical problems by addition of decimal fractions.
- Solve practical problems by subtraction of decimal fractions.
- Solve practical problems by combining addition and subtraction of decimal fractions.

Adding and subtracting decimal fractions are required at various stages in the design and manufacture of products. An estimator in the apparel industry adds and subtracts decimal fractions of an hour in finding cutting and sewing times. Most bakery cost and production calculations are expressed as decimal fractions. A salesclerk adds decimal fractions of dollars when computing sales checks.

ADDING DECIMAL FRACTIONS

To add decimal fractions, arrange the numbers so that the decimal points are directly under each other. The decimal point of a whole number is directly to the right of the last digit. Add each column as with whole numbers. Place the decimal point in the sum directly under the other decimal points.

> Example: Add. 8.75 + 231.062 + 0.739 8 + 0.007 + 23
>
> Arrange the numbers so that the decimal points are directly under each other.
>
> Add zeros so that all numbers have the same number of places to the right of the decimal point.
>
> Add each column of numbers.
>
> Place the decimal point in the sum directly under the other decimal points.
>
> $$\begin{array}{r} 8.750\,0 \\ 231.062\,0 \\ 0.739\,8 \\ 0.007\,0 \\ +\ 23.000\,0 \\ \hline 263.558\,8\ Ans \end{array}$$

142 Section 3 Decimal Fractions

Practice

To machine a swivel bracket, a model maker must know the length of stock required for the job. Find, in inches, the length of stock needed for the swivel bracket shown.

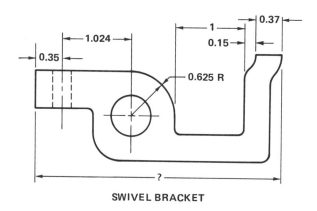

SWIVEL BRACKET

3.519" Ans

Exercise 15-1

Add the following numbers.

1. 0.237 + 0.395
2. 0.719 + 0.016 + 0.9
3. $\frac{1}{2}$ + 0.27 + 0.625
4. 0.836 + 2.917 + 0.02
5. 37.65 + $\frac{3}{4}$ + 0.133
6. 2 + 0.2 + 0.02 + 0.002
7. 132 + 0.06 + 0.7 + 3.8
8. 1.07 + 0.07 + 17.5 + $\frac{1}{8}$
9. 0.000 9 + 0.03 + 0.1 + 0.005
10. 12 + 0.057 + 9.6
11. 9.6 + 0.093 9 + 0.96
12. 0.012 + 0.007 5 + 303
13. 0.063 + 6$\frac{3}{10}$ + 630 + 0.63
14. 0.100 1 + 0.376 + 88
15. 837.14 + 19.3 + 0.187 + 0.007
16. 0.313 + 3.032 + 97$\frac{1}{40}$ + 0.138
17. 4.031 + 15.007 6 + 0.001 09
18. 232.032 + 23.203 2 + 0.232 032
19. 99.099 + $\frac{1}{500}$ + 0.06 + 22
20. 16.8 + 23.066 + 0.009 09 + 45

Add the following numbers. Round each sum to the indicated number of decimal places.

21. 0.084 + 0.998 8 (3 places)
22. 35.035 + 3 + $\frac{3}{4}$ (2 places)
23. 0.07 + 57.3 + 0.99 (1 place)

24. 3.6 + 0.006 6 + 29.86   (3 places)
25. 0.342 8 + 7.003 + 33.9   (1 place)
26. 301.43 + 30.143 + 0.301 43   (3 places)
27. 87.010 205 + 36$\frac{9}{64}$ (4 places)
28. 44.4 + 9.306 + 0.077 3   (2 places)

SUBTRACTING DECIMAL FRACTIONS

To subtract decimal fractions, arrange the numbers so that the decimal points are directly under each other. Subtract each column as with whole numbers. Place the decimal point in the difference directly under the other decimal points.

Examples:

1. Subtract. 44.6 − 27.368

   Arrange the numbers so that the decimal points are directly under each other. Add zeros so that the numbers have the same number of places to the right of the decimal point.

   $$\begin{array}{r} 44.600 \\ -\ 27.368 \\ \hline 17.232\ Ans \end{array}$$

   Subtract each column of numbers.

   Place the decimal point in the difference directly under the other decimal points.

2. Subtract. 136.307 9 − 87$\frac{3}{5}$

   Express the common fraction as a decimal fraction.

   $$\begin{array}{r} 136.307\ 9 \\ -\ \ 87.600\ 0 \\ \hline 48.707\ 9\ Ans \end{array}$$

   Arrange the numbers.

   Subtract.

   Place the decimal point in the difference directly under the other decimal points.

Section 3 Decimal Fractions

Practice

A brokerage clerk records the purchase price and sales price of stocks and bonds. Profit or loss on stocks and bonds are found as follows:

If the sales price is greater than the purchase price, the stock is sold at a profit.

If the sales price is less than the purchase price, the stock is sold at a loss.

The Record of Stock chart shown lists the sales prices and purchase prices of 3 firms. The prices are given in hundred dollars. A sales price of 26.37 is actually 26.37 hundred dollars or $2 637. Find the profit or loss for each firm.

RECORD OF STOCK (in hundred dollars)			
STOCK	FIRM A	FIRM B	FIRM C
Sales Price	26.37	9.3	17.68
Purchase Price	23.05	10.75	16.9
Profit or Loss			

Firm A: 3.32 hundred dollars or $332 Profit *Ans*
Firm B: 1.45 hundred dollars or $145 Loss *Ans*
Firm C: 0.78 hundred dollars or $78 Profit *Ans*

Exercise 15-2

Subtract the following numbers.

1. $7.932 - 3.107$
2. $2.005 - 0.222$
3. $0.98 - 0.899$
4. $0.001 - 0.000\ 1$
5. $18\frac{3}{8} - 16.027$
6. $383.3 - 294.16$
7. $45.05 - 44.999$
8. $0.9 - 0.000\ 9$
9. $16.018\ 1 - 0.796\ 03$
10. $0.414 - \frac{1}{4}$
11. $604.604 - 60.460\ 4$
12. $51.000\ 2 - 50.900\ 77$
13. $23.345 - 3.349\ 9$
14. $6\frac{91}{1\ 000} - 0.91$
15. $87.032 - 23.203\ 2$
16. $0.001\ 2 - 0.000\ 9$
17. $312.8 - 77.006\ 6$
18. $24.030\ 3 - 20\frac{3}{125}$
19. $9.406 - 0.014\ 3$
20. $0.000\ 1 - 0.000\ 01$

Subtract the following numbers. Round each answer to the indicated number of decimal places.

21. 0.708 4 − 0.606 (3 places)
22. 85.046 − 84.904 2 (3 places)
23. 41.005 3 − 39$\frac{7}{32}$ (4 places)
24. 0.036 − 0.009 (2 places)
25. 202.172 − 187.8 (1 place)
26. 504$\frac{24}{25}$ − 0.976 (2 places)
27. 0.200 46 − 0.200 459 (5 places)
28. 8.836 4 − 8.083 (3 places)

COMBINING ADDITION AND SUBTRACTION OF DECIMAL FRACTIONS

Often on-the-job computations require the combination of two or more different operations using decimal fractions. When solving a problem which requires both addition and subtraction operations, follow the procedures for each operation.

Example: A part of the structure of a building is shown. Girders are large beams under the first floor which carry the ends of the joists. Lally columns support the girders between the foundation walls. Find the length of the Lally column in inches.

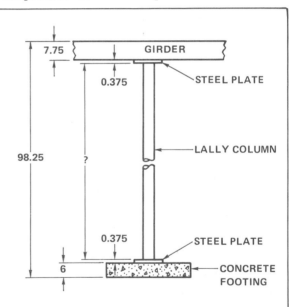

Length of Lally column = 98.250" − (6" + 0.375" + 0.375" + 7.750")

Add.
```
      6.000"
      0.375"
      0.375"
   +  7.750"
     14.500"
```

Subtract.
```
     98.250"
   − 14.500"
     83.750" Ans
```

Section 3 Decimal Fractions

UNIT REVIEW

Exercise 15-3

Add the following numbers.

1. 0.413 + 0.033
2. 0.305 + 0.106 + 0.4
3. $\frac{5}{16}$ + 0.080 8 + 0.590 9
4. 16.016 + 29 + 0.8
5. 0.000 3 + 0.003 + 0.03
6. 2.12 + 6.048 + 0.811
7. 77.77 + 0.311 08 + 66
8. 0.000 1 + $\frac{1}{80}$ + 9.034

9. 342.083 8 + 61 + 0.730 12
10. 0.021 + 0.008 2 + 606
11. 0.002 + 0.000 2 + 0.000 02
12. $77\frac{7}{25}$ + 4.031 + 0.8 + 6.081
13. 100 + 0.005 + 0.393 + 5.808 7
14. 51.006 7 + 21 + 21.021 + 2.102 1
15. 494.206 3 + 90.631 + 0.241 6
16. 9.3 + 0.93 + 0.093 + 0.009 3

Subtract the following numbers.

17. 0.783 − 0.678
18. 1.031 − 0.744
19. 0.95 − 0.304 2
20. 0.002 − 0.000 9
21. $36\frac{1}{8}$ − 36.124
22. 26.081 8 − 12.603
23. 15.100 2 − 14.900

24. 8.845 − 8.844 9
25. 71.071 − $68\frac{197}{200}$
26. 294.66 − 294.067 3
27. 0.000 9 − 0.000 89
28. 86.875 3 − 19.19
29. $\frac{1}{250}$ − 0.003 8
30. 6.000 2 − 5.808 06

Refer to the chart for problems 31–36. Find how much greater value **A** is than value **B**.

	A	B
31.	312.067 pounds + 84.12 pounds	107.34 pounds + 172.9 pounds
32.	45.18 metres + 16.25 metres	55.055 metres + 3.25 metres
33.	673.2 litres + 107.67 litres	703.98 litres + 12.6 litres
34.	9.5 inches + 14.66 inches	16.37 inches + 0.878 inch
35.	3.3 weeks + 0.87 week	1.36 weeks + 1.08 weeks
36.	6.125 kilometres + 4.250 kilometres	7.02 kilometres + 2.925 kilometres

Unit 15 Addition and Subtraction of Decimal Fractions 147

PRACTICAL APPLICATIONS

Exercise 15-4

1. A hardware store clerk bills a customer for the following items: nails, $6.85; locks, $13.47; hinges, $5.72; drawer pulls, $4; and cabinet catches, $6.09. What is the total amount of the bill?

2. Find, in centimetres, dimensions **A–F** of the profile gauge shown.

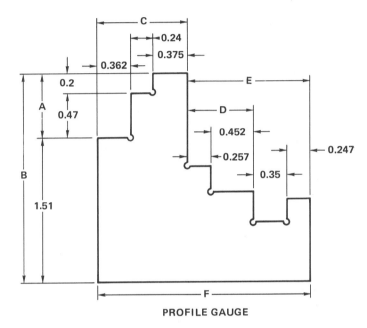

PROFILE GAUGE

3. In the design of products, it is important to consider the weights of the materials of which the products are made.

	Aluminum	Copper	Tin	Brass
Weight Per Cubic Foot	168.5 lb	554.7 lb	454.9 lb	536.6 lb
Weight Per Cubic Inch	0.097 5 lb	0.321 0 lb	0.263 3 lb	0.310 5 lb

a. Determine the difference in weight of 1 cubic foot of each of the following:
   (1) aluminum and tin
   (2) tin and copper
   (3) brass and copper
   (4) copper and aluminum

b. Determine the difference in weight of 1 cubic inch of each of the following:
   (1) brass and aluminum
   (2) copper and brass
   (3) tin and brass
   (4) aluminum and tin

148   Section 3   Decimal Fractions

4. A thickness or feeler gauge is shown. Thickness gauges are widely used in automotive, aviation, manufacturing, and machine service and repair occupations. An engine repair technician uses a thickness gauge to adjust tappets, spark plugs, and distributor points. Bearing clearances and gear play are checked and pistons, rings, and pins are fitted using thickness gauges. Find the smallest combination of gauge leaves which total each of the following thicknesses: (More than one combination may total certain thicknesses.)

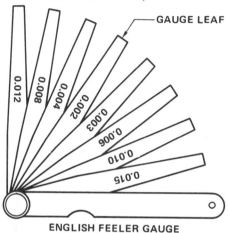

a. 0.014"   c. 0.021"   e. 0.011"   g. 0.029"
b. 0.033"   d. 0.038"   f. 0.042"   h. 0.049"

5. The following number of cubic metres of concrete are delivered to a construction site in one week: 20.5, 32.8, 18, 28.75, and 48.3 cubic metres. How many total cubic metres are delivered during the week?

6. Find in inches, each of the following distances on the base plate shown.

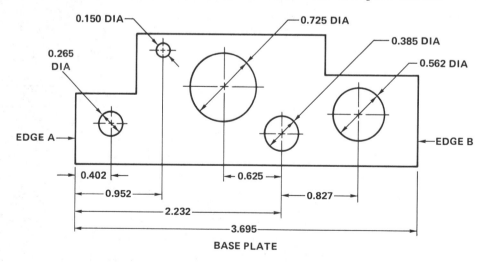

a. The center distance between the 0.265" diameter hole and the 0.150" diameter hole.
b. The center distance between the 0.385" diameter hole and the 0.150" diameter hole.
c. The distance between edge A and the center of the 0.725" diameter hole.
d. The distance between edge B and the center of the 0.385" diameter hole.
e. The distance between edge B and the center of the 0.562" diameter hole.

7. A hex bolt, lockwasher, and nut are used to fasten 3 pieces of ground steel.

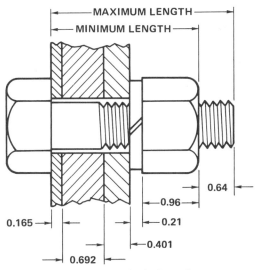

a. Find, in centimetres, the minimum bolt lengths.
b. Find, in centimetres, the maximum bolt lengths.

8. An automatic screw machine supervisor estimated the set-up times for 4 different jobs at a total of 6.25 hours. The jobs actually took 1.75 hours, 0.6 hour, 2.125 hours, and 1.4 hours respectively. By how many hours was the total of the 4 jobs overestimated?

9. During a one-year period, the following kilowatt-hours (kW·h) of electricity were consumed in a home.

January   693.75 kW·h   May   663.18 kW·h   September 668.43 kW·h
February 678.24 kW·h   June   658.33 kW·h   October   659.98 kW·h
March   674.83 kW·h   July   665.09 kW·h   November 671.06 kW·h
April   666.05 kW·h   August   672.46 kW·h   December 682.33 kW·h

a. Find the total number of kilowatt-hours of electricity consumed for the year.
b. How many more kilowatt-hours of electricity are used during the first four months than the second four months of the year?
c. How many more kilowatt-hours of electricity are used during the highest monthly consumption than during the lowest monthly consumption?

10. When overhauling an engine, an automobile mechanic grinds the cylinder walls. After grinding, larger diameter pistons than the original pistons are installed. On a certain job, the mechanic grinds the cylinder walls which increases the diameter of the cylinders by 0.040 inch. The diameters of the cylinders before grinding are 3.625 inches. A clearance of 0.002 5″ is required between the piston and cylinder wall. What size (diameter) pistons are ordered for this job?

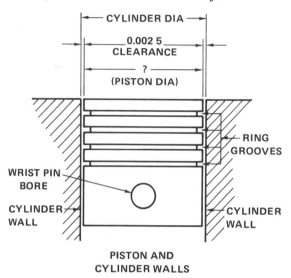

PISTON AND
CYLINDER WALLS

11. Find, in millimetres, dimensions **A**, **B**, **C**, and **D** of the idler bracket shown.

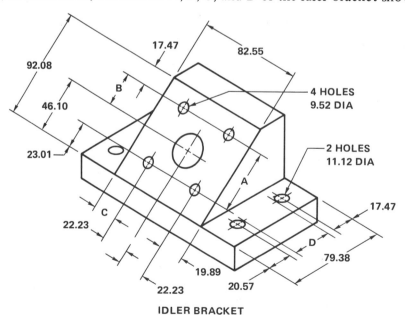

IDLER BRACKET

12. An environmental systems technician finds the volume of air flowing through pipes. The amount (volume) of air flowing through a pipe depends on the velocity of the air and the size (diameter) of the pipe. The table lists the number of cubic feet of air flowing at various velocities through different diameter pipes.

Velocity of Air in feet per second	Inside Diameter of Pipe in Inches			
	1" Dia	2" Dia	6" Dia	10" Dia
2 ft/s	0.65 cu ft/min	2.62 cu ft/min	23.6 cu ft/min	65.4 cu ft/min
5 ft/s	1.64 cu ft/min	6.55 cu ft/min	59.0 cu ft/min	163.0 cu ft/min
8 ft/s	2.62 cu ft/min	10.50 cu ft/min	94.0 cu ft/min	262.0 cu ft/min
12 ft/s	3.93 cu ft/min	15.7 cu ft/min	141.0 cu ft/min	393.0 cu ft/min

a. At a velocity of 2 ft/s what is the total volume of air per minute which flows through four pipes with diameters of 1", 2", 6", and 10"?

b. In one minute how much more air flows through a 6" diameter pipe at 5 ft/s than a 2" diameter pipe at 8 ft/s?

c. In one minute how much more air flows through a 10" diameter pipe at 5 ft/s than a 6" diameter pipe at 12 ft/s?

d. In one minute how much more air flows through three 1" diameter pipes at 12 ft/s than two 2" diameter pipes at 2 ft/s?

e. In one minute how much more air flows through one 10" diameter pipe at 2 ft/s than three 2" diameter pipes at 5 ft/s?

UNIT 16
Multiplication of Decimal Fractions

OBJECTIVES

After studying this unit you should be able to
- Multiply decimal fractions.
- Solve practical problems by multiplication of decimal fractions.
- Solve practical problems by combining addition, subtraction, and multiplication of decimal fractions.

A payroll clerk computes the weekly wage of an employee who works 36.25 hours at an hourly rate of $6.08. In preparing a solution, a laboratory technician computes 0.125 of 0.5 litre of acid. A chef finds the total food cost for a banquet of 46 persons at $6.75 per person. A homeowner checks an electricity bill of $33.80 for 650 kilowatt-hours of electricity at $0.052 per kilowatt-hour. Multiplication of decimal fractions is required for these computations.

MEANING OF MULTIPLICATION OF DECIMALS

Just as with whole numbers and common fractions, multiplication of decimal fractions is a short method of adding equal amounts. For example, 4 × 0.3 means 0.3 is taken 4 times.

$$0.3 + 0.3 + 0.3 + 0.3$$

Adding the decimal fractions gives the sum 1.2.

$$4 \times 0.3 = 1.2$$

The enlarged decimal inch scale shows an inch divided into 10 smaller equal units. Each small unit equals 0.1 inch.

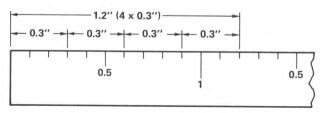

ENLARGED DECIMAL INCH SCALE

Unit 16 Multiplication of Decimal Fractions

Recall that multiplication of whole numbers and fractions can be done in any order. Multiplication of decimal fractions can also be done in any order. For example, 4 × 0.3 equals 0.3 × 4.

MULTIPLYING DECIMAL FRACTIONS

To multiply decimal fractions, multiply using the same procedure as with whole numbers. Count the number of decimal places in both the multiplier and multiplicand. Begin counting from the last digit on the right of the product and place the decimal point the same number of places as there are in both the multiplicand and the multiplier.

Example: Multiply. 60.412 × 0.53

Multiply as with whole numbers.

Beginning at the right of the product, place the decimal point the same number of decimal places as there are in both the multiplicand and the multiplier.

$$\begin{array}{r} 60.412 \text{ (3 places)} \\ \times \quad 0.53 \text{ (2 places)} \\ \hline 1\ 81236 \\ 30\ 2060 \\ \hline 32.01836 \text{ (5 places)} \end{array}$$

32.018 36 *Ans*

When multiplying certain decimal fractions, the product has a smaller number of digits than the number of decimal places required. For these products, add as many zeros to the left of the product as are necessary to give the required number of decimal places.

Example: Multiply. 0.004 7 × 0.08. Round the answer to four decimal places.

Multiply as with whole numbers.

The product must have 6 decimal places.
Add three zeros to the left of the product.

$$\begin{array}{r} 0.0047 \text{ (4 places)} \\ \times \quad 0.08 \text{ (2 places)} \\ \hline 0.000376 \text{ (6 places)} \end{array}$$

Round to 4 decimal places.

0.000 4 *Ans*

Practice

A drill press operator is required to drill 35 equally spaced holes in a piece of strip stock shown. Find, in centimetres, the distance between hole #1 and hole #35.

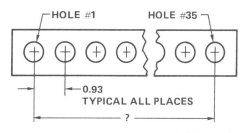

31.62 cm *Ans*

Exercise 16-1

Multiply the following numbers.

1. 0.6 × 0.9
2. 0.3 × 22
3. 0.42 × 0.8
4. 0.875 × 12
5. 0.44 × 5.7
6. 10.25 × 0.12
7. 22.22 × $\frac{3}{4}$
8. 0.053 × 0.4
9. 4.001 × 9.8
10. 0.314 16 × 6
11. 6.36 × 7.3
12. 0.029 × 0.05
13. $3\frac{7}{8}$ × 3.66
14. 0.001 × 0.01
15. 81.913 × 4.08
16. 0.012 5 × 0.246
17. 124 × 4.001 3
18. 0.008 × 0.019
19. 5.077 × $6\frac{3}{25}$
20. 0.000 2 × 104

Multiply the following numbers. Round the answers to the indicated number of decimal places.

21. 0.8 × 0.4   (1 place)
22. 0.009 × 0.5   (3 places)
23. 3.18 × 6.736   (4 places)
24. 800.75 × 10.1   (2 places)
25. 1.08 × $2\frac{9}{16}$ (3 places)
26. 0.030 4 × 0.088   (5 places)
27. 929.29 × 551.24   (2 places)
28. 0.001 × 0.006   (5 places)

MULTIPLYING THREE OR MORE FACTORS

When multiplying three or more factors, multiply two factors. Then multiply the product by the third factor. Continue the process until all factors are multiplied.

Example: Find the product. 0.74 × 14 × 3.8 × $\frac{3}{5}$

Multiply.	0.74 × 14 = 10.36
Multiply.	10.36 × 3.8 = 39.368
Express the common fraction as a decimal fraction and multiply.	39.368 × 0.6 = **23.620 8** *Ans*

Practice

Large metal parts can be electroplated in a plating bath tank. A plating bath is a solution which contains the plating metal. Parts to be plated are immersed in the bath for a certain period of time. Electroplating provides corrosion protection and often makes a product more attractive. An electroplater finds the number of gallons of plating bath needed for various tank sizes. Find the number of gallons of plating bath needed to provide a 4 1/4-foot bath depth in the tank shown. One cubic foot of liquid contains 7.479 gallons. Round the answer to the nearer tenth gallon.

$V = lwh$ where V = volume
l = length
w = width
h = height

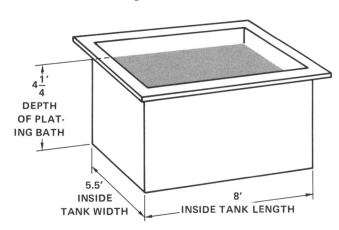

1 398.6 gallons *Ans*

Exercise 16-2

Multiply the following numbers.

1. 0.74 × 12 × 0.5
2. 0.009 × 0.09 × 0.9
3. 1.7 × $\frac{3}{8}$ × 61.7
4. 0.031 × 0.825 × 10
5. 0.5 × 0.5 × 5.5 × 55
6. 2.54 × 3.14 × 2 × 0.03
7. 0.81 × $\frac{21}{50}$ × 0.6 × 17
8. 0.001 × 1 000 × 0.01 × 100
9. 43.93 × 7.07 × 6 × 0.9
10. 0.012 5 × 0.02 × 41.8
11. 94.808 × 3.12 × 2 × 0.1
12. 2.383 × 4 × 0.99
13. 3.3 × 0.33 × $\frac{33}{1\ 000}$ × 33
14. 2 817 × 0.63 × 78.007

Multiply the following numbers. Round the answers to the indicated number of decimal places.

15. 0.87 × 3.12 × 0.06 (4 places)
16. 129.8 × 0.077 × $\frac{1}{2}$ × 16.32 (3 places)

17. 14.9 × 8.25 × 105 × 0.4   (1 place)
18. 0.021 × 0.376 × 0.6 × 42   (5 places)
19. 88.99 × 5.4 × 45 × 0.46   (3 places)

MULTIPLYING BY POWERS OF TEN

The method of multiplying by powers of ten is quick and easy to apply. The decimal system is based on groupings of ten. This method of multiplication is based on groupings of ten place values.

To multiply a number by 10, 100, 1 000, 10 000, etc., move the decimal point in the multiplicand as many places to the right as there are zeros in the multiplier. If there are not enough digits in the multiplicand, add zeros to the right of the multiplicand.

Examples:
1. 0.085 × 10 = 0.85 *Ans*   (1 zero in 10; move 1 place to the right, 0.0 85)
2. 0.085 × 100 = 8.5 *Ans*   (2 zeros in 100; move 2 places to the right, 0.08 5)
3. 0.085 × 1 000 = 85 *Ans*   (3 zeros in 1 000; move 3 places to the right, 0.085)
4. 0.085 × 10 000 = 850 *Ans*   (4 zeros in 10 000; move 4 places to the right, 0.0850)
   It is necessary to add 1 zero to the right since the multiplicand 0.085 has only 3 digits.

To multiply a number by 0.1, 0.01, 0.001, 0.000 1, etc., move the decimal point in the multiplicand as many places to the left as there are decimal places in the multiplier. If there are not enough digits in the multiplicand, add zeros to the left of the multiplicand.

Examples:
1. 127.5 × 0.1 = 12.75 *Ans*   (1 decimal place in 0.1; move 1 place to the left, 12 7.5)
2. 127.5 × 0.01 = 1.275 *Ans*   (2 decimal places in 0.01; move 2 places to the left, 1 27.5)
3. 127.5 × 0.001 = 0.127 5 *Ans*   (3 decimal places in 0.001; move 3 places to the left, 0 127.5)
4. 127.5 × 0.000 1 = 0.012 75 *Ans*   (4 decimal places in 0.000 1; move 4 places to the left, 0 0127.5)
   It is necessary to add 1 zero to the left since the multiplicand 127.5 has only 3 digits to the left of the decimal point.

Exercise 16-3

Multiply the following numbers. Use the rules for multiplying by a power of ten.

1. 0.72 × 10
2. 0.072 × 10
3. 0.912 × 100
4. 18.7 × 1 000
5. 1.87 × 1 000
6. 0.005 × 100
7. 0.039 × 10 000
8. 312.88 × 100 000
9. 3.7 × 0.1
10. 3.7 × 0.01
11. 0.08 × 0.01
12. 25.032 × 0.001
13. 843 × 0.001
14. 843 × 0.000 1
15. 63 302 × 0.000 01
16. 900.3 × 0.000 1
17. 0.033 × 100
18. 23.4 × 0.01
19. 9.35 × 1 000
20. 23.4 × 0.001
21. 0.002 × 100
22. 0.072 3 × 10 000
23. 8 932 × 0.001
24. 707 × 0.000 1

COMBINING ADDITION, SUBTRACTION, AND MULTIPLICATION OF DECIMAL FRACTIONS

When solving a problem which requires two or more different operations, think the problem through to determine the steps to be used in the solution. Then follow the procedures for each operation.

> Example: A payroll clerk determines employee wages by finding the gross wages and deductions. A certain hourly paid employee is paid $5.38 per hour for 40 hours. Time-and-a-half is paid for any hours worked over the regular 40 hour workweek. The employee worked 42.75 hours during one week. The total deductions for the week are $58.51. Find the net wage (take-home pay) of the employee.
>
> | | |
> |---|---|
> | Regular wage for 40 hours. | 40 × $5.38 = $215.20 |
> | Overtime hours. | 42.75 hours − 40 hours = 2.75 hours |
> | Overtime wage. | 1.5 × 2.75 hours × $5.38/hour = $22.19 |
> | Gross wage. | $215.20 + $22.19 = $237.39 |
> | Net wage. | $237.39 − $58.51 = $178.88 *Ans* |

UNIT REVIEW

Exercise 16-4

Multiply the following numbers.

1. 0.8 × 0.7
2. 0.4 × 36
3. 0.57 × 0.5
4. 18.13 × 0.14
5. 62.28 × $\frac{1}{4}$
6. 0.077 × 0.9
7. 6.002 × 3.3
8. 0.024 × 0.06
9. $4\frac{5}{8}$ × 4.32
10. 73.881 × 1.08
11. 0.013 × 0.014
12. 67.022 × 0.038
13. $3\frac{3}{10}$ × 4.078
14. 74.03 × 0.009

Multiply the following numbers. Round the answers to the indicated number of decimal places.

15. 812.23 × 11.7   (2 decimal places)
16. 6.06 × $5\frac{3}{16}$ (3 decimal places)
17. 0.008 × 0.005 2   (5 decimal places)

Multiply the following numbers.

18. 0.85 × 14 × 0.7
19. 0.007 × 0.07 × 0.7
20. 2.8 × $\frac{3}{5}$ × 18.8
21. 0.014 × 0.913 × 12
22. 3.45 × 4.31 × 3 × 0.04
23. 32.3 × 6.06 × 5 × 0.2
24. 0.013 × 0.09 × 32.2
25. 88.909 × 2.11 × 2 × 0.8
26. 4.3 × 0.43 × $\frac{43}{1\,000}$ × 43
27. 891 × 0.77 × 66.005

Multiply the following numbers. Round the answers to the indicated number of decimal places.

28. 0.79 × 8.05 × 0.07   (4 decimal places)
29. 13.3 × 7.34 × 0.301 × 0.9   (2 decimal places)
30. 218.6 × 0.89 × $\frac{9}{10}$ × 27.51   (3 decimal places)

Multiply the following numbers. Use the rules for multiplying by a power of ten.

31. 0.81 × 10
32. 0.997 × 100
33. 16.3 × 1 000
34. 0.003 × 100

35. 662.11 × 10 000
36. 9.7 × 0.1
37. 6.5 × 0.01
38. 13.051 × 0.001
39. 18 733 × 0.000 01

40. 512.6 × 0.000 1
41. 0.44 × 100
42. 32.8 × 0.01
43. 0.003 × 1 000
44. 7 804 × 0.001

Refer to the chart for 45–50. Find how much greater value **A** is than value **B**.

	A	B
45.	0.67 × 0.83 kilogram	0.42 × 0.63 kilogram
46.	1.04 × 27 inches	0.99 × 26 inches
47.	0.2 × 59.75 hours	0.3 × 36.25 hours
48.	23.05 × 19 metres	19.8 × 17 metres
49.	4.8 × 20.5 gallons	5.2 × 12.3 gallons
50.	0.6 × 88.3 miles	0.4 × 79.9 miles

PRACTICAL APPLICATIONS

Exercise 16-5

1. Spur gears are used to transmit power between parallel shafts. Gear designers and gear cutters work to a high degree of precision. Ten-thousandths inch precision is usually required. Circular pitch, working depth, clearance, and tooth thickness are basic dimensions which are used. Find, to four decimal places, the working depths, clearances, and tooth thicknesses shown in the table.

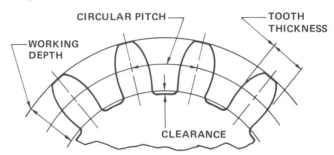

Working depth = 0.636 6 × Circular Pitch
Clearance = 0.05 × Circular Pitch
Tooth thickness = 0.5 × Circular Pitch

	Circular Pitch	Working Depth	Clearance	Tooth Thickness
a.	0.392 5″			
b.	0.158 2″			
c.	0.875 9″			
d.	1.237 8″			
e.	1.593 1″			

2. A drygoods salesperson takes the following order for three pieces of fabric: 2 1/4 yards at $2.27 per yard, 3 3/4 yards at $1.96 a yard, and 4 1/2 yards at $2.38 a yard. Sales tax is $0.92. Find the total price charged the customer for this order.

3. A mason loads materials for a job on a pickup truck. The truck is rated to carry a maximum load of ·1.75 tons (1 short ton = 2 000 lb). The following materials are loaded on the truck: 115 four-foot lengths of reinforcing rod which weigh 0.375 lb/ft, 110 hollow wall tiles which weigh 21.25 pounds each, and 8 1/2 bags of mortar which weigh 100 pounds each. How many more pounds of materials can be loaded on the truck to bring the complete load to 1.75 tons?

4. A nursery owner raises and sells plants. To insure the best plant growth, an acid plant food is fed to evergreens. The plant food is labeled 30-10-10. It gives the amount of each of primary ingredients of the total plant food mixture. In this mixture 30/100 or 0.30 of the total mixture is nitrogen, 10/100 or 0.10 is phosphoric acid, and 10/100 or 0.10 is potash. The plant food is available in 3 different size containers. Find the weights of each ingredient in each container.

	POUNDS OF PLANT FOOD	POUNDS OF NITROGEN	POUNDS OF PHOSPHORIC ACID	POUNDS OF POTASH
a.	1.25 lb			
b.	3.50 lb			
c.	10.75 lb			

5. An inspector measures various dimensions of a part. Using the front view of the mounting block, find dimensions **A**, **B**, and **C** in millimetres.

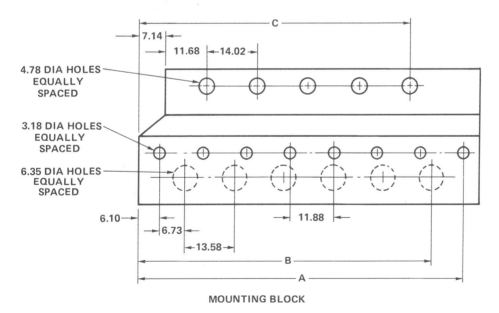

MOUNTING BLOCK

Note: The 6.35 mm diameter holes are drawn with broken or hidden lines. Broken lines show that the holes are drilled from the back and do not go through the part. The holes cannot be seen when viewing the part from the front.

6. An air conditioning technician determines ventilation air requirements of a building by either of the following two methods:

- Total ventilation air required in cubic feet per minute = ventilation required per person in cubic feet per minute × number of persons.

- Total ventilation required in cubic feet per minute = ventilation required per square foot of floor in cubic feet per minute × number of square feet of floor.

The ventilation requirements depend on the use of the building. The ventilation requirements are greater for a hospital than for a store having the same number of persons or the same number of square feet. The table lists different types of buildings with their per person and per square foot of floor requirements.

TYPE OF BUILDING	VENTILATION AIR REQUIREMENTS IN CUBIC FEET PER MINUTE	
	Per Person	Per Square Foot of Floor
Apartment	17.5	0.33
Store	6.25	0.05
Factory	8.75	0.10
Hospital	25.5	0.33
Office	20.25	0.25

What is the total ventilation air required in cubic feet per minute for each of the following buildings?
a. An apartment with 1 250 square feet.
b. A factory with 115 employees.
c. A factory with 12 500 square feet.
d. An office with 15 employees.
e. An apartment with 5 occupants.
f. A hospital with 23 650 square feet.
g. A supermarket with 11 800 square feet.
h. A hospital with an average of 210 patients and 85 employees.

7. A high performance small block 8 cylinder engine is designed to produce 1.4 brake horsepower for each cubic inch of piston displacement. Each piston displaces 32.33 cubic inches. Find the total brake horsepower of the engine.

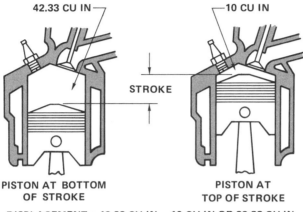

DISPLACEMENT = 42.33 CU IN − 10 CU IN OR 32.33 CU IN

8. In order to determine selling prices of products, a baker finds the cost of ingredients and adds estimated profit. For large production products, ingredient costs are often broken down to three-place decimal fractions of a dollar per ounce of ingredients. The total ingredient cost for a cake is $0.065 per ounce (1 pound = 16 ounces). The estimated profit is $0.73. Find the selling price of a cake which weighs 1.625 pounds.

9. A payroll clerk computes the net wages for employees A, B, and C listed in the table. The regular workweek is 40 hours with time-and-a-half paid for any hours over 40. Use 0.061 3 and 0.02 for social security and retirement deductions. Determine the net wage for each employee.

Employee	Number of Hours Worked in Week	Hourly Rate of Pay	Withholding Tax	Health and Accident Insurance
A	44.25 h	$5.43	$36.27	$5.53
B	37.75 h	$5.87	$35.08	$4.92
C	46.5 h	$5.12	$37.11	$5.16

10. Materials expand when heated. Different materials expand at different rates. Mechanical and construction technicians must often consider material rates of expansion. The amount of expansion for short lengths of material is very small. Expansion is computed if a product is made to a high degree of precision and is subjected to a large temperature change. Also, expansion is computed for large structures because of the long lengths of structural members used. The table lists the expansion per inch for each Fahrenheit degree rise in temperature for a 1-inch length of material.

Material	Expansion of 1 inch Length of Material in 1 Fahrenheit Degree Temperature Increase
Aluminum	0.000 012 44 inch
Copper	0.000 009 00 inch
Structural Steel	0.000 007 22 inch
Brick	0.000 003 0 inch
Concrete	0.000 008 0 inch

Find the total expansion for each of the materials listed. Express the answer to 3 decimal places.

	Material	Length Before Temperature Increase	Fahrenheit Temperature Change	Total Expansion in Inches
a.	Copper Cable	300 ft	From 35°F to 92.5°F	
b.	Steel I Beam	50 ft	From 5.3°F to 90°F	
c.	Brick Wall	225 ft	From 20.8°F to 81°F	
d.	Aluminum Wire	510 ft	From 43°F to 105.7°F	
e.	Concrete Foundation	150 ft	From 15.2°F to 89.6°F	

UNIT 17
Division of Decimal Fractions

OBJECTIVES

After studying this unit you should be able to

- Divide decimal fractions.
- Solve practical problems by division of decimal fractions.
- Solve practical problems by combining division with one or more additional arithmetic operation.

A retailer divides with decimal fractions when computing the unit cost of a product purchased in wholesale quantities. Insurance rates and claim payments are calculated by division of decimal fractions. Division with decimal fractions is used to compute manufacturing time per piece after total production times are determined.

MEANING OF DIVISION OF DECIMAL FRACTIONS

As with division of whole numbers and common fractions, division of decimal fractions is a short method of subtracting a subtrahend a given number of times. It is important that you understand the meaning of division of decimal fractions.

Dividing 0.8 by 0.2 means 0.2 is subtracted from 0.8 until 0.2 can no longer be subtracted.

$$0.8 - 0.2 = 0.6$$
$$0.6 - 0.2 = 0.4$$
$$0.4 - 0.2 = 0.2$$
$$0.2 - 0.2 = 0$$

Since 0.2 is subtracted 4 times from 0.8, there are four 0.2's contained in 0.8.

$$0.8 \div 0.2 = 4$$

166 Section 3 Decimal Fractions

Another way of developing an understanding of division of decimal fractions is to locate the values on the enlarged inch scale. Note each small division equals 0.1″. The enlarged inch scale shows four 0.2 parts in 0.8 inch.

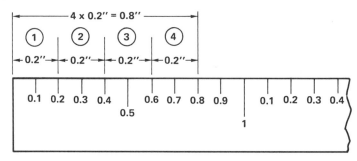

ENLARGED INCH SCALE

DIVIDING DECIMAL FRACTIONS

To divide decimal fractions use the same procedure as with whole numbers. Move the decimal point of the divisor as many places to the right as necessary to make the divisor a whole number. Move the decimal point of the dividend the same number of places to the right. Since division can be expressed as a fraction the value does not change if both the numerator and denominator are multiplied by the same number. Add zeros to the dividend if necessary. Place the decimal point in the quotient directly above the decimal point in the dividend. Divide as with whole numbers. Zeros may be added to the dividend to give the degree of precision needed in the quotient.

Examples:

1. Divide. 0.338 0 ÷ 0.52

    Move the decimal point 2 places to the right in the divisor.

    Move the decimal point 2 places in the dividend.

    Place the decimal point in the quotient directly above the decimal point in the dividend.

    Divide.

    $$\begin{array}{r} 0.65 \; Ans \\ 0\,52.\,\overline{)\,0\,33.80} \\ \underline{31\;2} \\ 2\;60 \\ \underline{2\;60} \end{array}$$

2. Divide. 11.9 ÷ 3.072
   Round the answer to 3 decimal places.

   Move the decimal point 3 places to the right in the divisor.

   Move the decimal point 3 places in the dividend, adding two zeros.

   Place the decimal point directly above the decimal point in the dividend.

   Add 4 zeros to the dividend. One more zero is added than the degree of precision required.

```
                3.873 6 ≈ 3.874 Ans
3 072. ) 11 900.000 0
         9 216
         2 684 0
         2 457 6
           226 40
           215 04
            11 360
             9 216
             2 144 0
             1 843 2
               300 8
```

Practice

1. A structural column which has a cross-sectional area of 3.188 square inches supports a floor load of 5 1/8 tons. Find the load to 3 decimal places on 1 square inch of cross-sectional area.

    1.608 tons *Ans*

2. Twenty grooves are machined in a plate. All grooves are equally spaced. Find the center-to-center distance, dimension A, between two consecutive grooves.

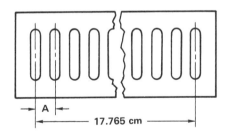

0.935 cm *Ans*

Exercise 17-1

Divide the following numbers.

1. 0.6 ÷ 0.3
2. 1.2 ÷ 0.4
3. 0.8 ÷ 0.02
4. 12.3 ÷ 4.1

5. $0.72 \div 0.04$
6. $36.750 \div \frac{1}{4}$
7. $0.367\ 5 \div \frac{1}{4}$
8. $8.024 \div 1.003$
9. $48.036 \div 12$
10. $105 \div 5.25$
11. $10.5 \div 52.5$
12. $0.001 \div 0.125$
13. $0.187\ 5 \div 1.5$
14. $35\frac{15}{16} \div 7.187\ 5$
15. $0.756 \div 0.3$
16. $5.948 \div 14.87$
17. $5\ 948 \div 148.7$
18. $292.114\ 2 \div 18.606$
19. $1.987\ 5 \div \frac{5}{16}$
20. $0.000\ 065\ 1 \div 0.009\ 3$

Divide the following numbers. Round the answers to the indicated number of decimal places.

21. $0.613 \div 0.912$ (3 places)
22. $7.059 \div 6.877$ (2 places)
23. $28\ 900 \div 440\ 110$ (4 places)
24. $6.998 \div 0.03$ (2 places)
25. $0.423\ 3 \div \frac{3}{4}$ (3 places)
26. $5\ 189.7 \div 6.9$ (1 place)
27. $0.682 \div 67.1$ (4 places)
28. $9.102\ 08 \div \frac{7}{32}$ (5 places)
29. $0.009\ 3 \div 0.979$ (2 places)
30. $15.630\ 9 \div 1.842$ (3 places)

DIVIDING BY POWERS OF TEN

Division is the inverse of multiplication. Dividing by 10 is the same as multiplying by 1/10 or 0.1. For example $32.5 \div 10 = 32.5 \times 0.1$. Dividing a number by 10, 100, 1 000, 10 000, etc., is the same as multiplying a number by 0.1, 0.01, 0.001, 0.000 1, etc. To divide a number by 10, 100, 1 000, 10 000, etc., move the decimal point in the dividend as many places to the left as there are zeros in the divisor. If there are not enough digits in the dividend, add zeros to the left of the dividend.

Examples:
1. $732.4 \div 10 = 73.24$ *Ans* (1 zero in 10; move 1 place to the left, 73_2.4)
2. $732.4 \div 100 = 7.324$ *Ans* (2 zeros in 100; move 2 places to the left, 7_32.4)
3. $732.4 \div 1\ 000 = 0.732\ 4$ *Ans* (3 zeros in 1 000; move 3 places to the left, 0_732.4)
4. $732.4 \div 10\ 000 = 0.073\ 24$ *Ans* (4 zeros in 10 000; move 4 places to the left, 0_0732.4)
   It is necessary to add 1 zero to the left since the dividend, 732.4, has only 3 digits to the left of the decimal point.

Dividing a number by 0.1, 0.01, 0.001, 0.000 1, etc., is the same as multiplying by 10, 100, 1 000, 10 000, etc. To divide a number by 0.1, 0.01, 0.001, 0.000 1, etc., move the decimal point in the dividend as many places to the right as there are decimal places in the divisor. If there are not enough digits in the dividend, add zeros to the right of the dividend.

Examples:
1. 0.065 ÷ 0.1 = 0.65 *Ans*   (1 decimal place in 0.1; move 1 place to the right, 0.0 65)
2. 0.065 ÷ 0.01 = 6.5 *Ans*   (2 decimal places in 0.01; move 2 places to the right, 0.06 5)
3. 0.065 ÷ 0.001 = 65 *Ans*   (3 decimal places in 0.001; move 3 places to the right, 0.065)
4. 0.065 ÷ 0.000 1 = 650 *Ans*   (4 decimal places in 0.000 1; move 4 places to the right, 0.0650)

It is necessary to add 1 zero to the right since the dividend, 0.065, has only 3 digits.

Practice

1. Cable is purchased from an electrical supplier at $172 per thousand feet. Find the cost per foot.

    $0.172/ft *Ans*

2. A technical illustrator must know the number of 0.1-centimetre spaces that can be laid out in an 8.5-centimetre length.

    85 *Ans*

Exercise 17-2

Divide the following numbers. Use the rules for dividing by a power of ten.

1. 0.37 ÷ 10
2. 6.89 ÷ 100
3. 0.09 ÷ 100
4. 36 051 ÷ 1 000
5. 732 ÷ 1 000
6. 917 ÷ 10 000
7. 72 086 ÷ 100 000
8. 806.58 ÷ 10 000
9. 0.88 ÷ 0.1
10. 0.054 ÷ 0.1
11. 0.901 ÷ 0.01
12. 17.7 ÷ 0.001

13. $1.77 \div 0.001$
14. $0.005 \div 0.01$
15. $0.052 \div 0.000\ 1$
16. $3.288 \div 0.000\ 01$
17. $0.044 \div 0.1$
18. $28.2 \div 100$
19. $9.08 \div 0.001$
20. $32.7 \div 1\ 000$
21. $0.005 \div 0.01$
22. $0.027\ 3 \div 0.000\ 1$
23. $7\ 831 \div 1\ 000$
24. $818 \div 10\ 000$

COMBINING ADDITION, SUBTRACTION, MULTIPLICATION AND DIVISION OF DECIMAL FRACTIONS

Actual on-the-job problems often require a combination of different operations in their solutions. A problem must first be thought through to determine how it is going to be solved. After the steps in the solution are determined, apply the procedures for each operation.

Example: An estimator for a die-casting firm quotes an order for 2 850 castings at a selling price of $3.43 per casting. The materials (molten metals) used to produce the 2 850 castings are listed in the table. In addition to materials, costs in producing the 2 850 castings are as follows: die cost, $1 750.75; overhead cost, $510.25; labor cost, $625.50.

MATERIAL REQUIRED TO PRODUCE 2 850 CASTINGS		
Material (Molten Metal)	Number of Pounds Required	Cost Per Pound
Zinc	5 520	$0.52
Aluminum	308.5	$0.43
Copper	180.25	$0.78

Find the profit per casting.

Material cost

Cost of zinc	$5\ 520 \times \$0.52 = \$2\ 870.40$
Cost of aluminum	$308.5 \times \$0.43 = \$132.66$
Cost of copper	$180.25 \times \$0.78 = \$140.60$
Cost of materials	$\$2\ 870.40 + \$132.66 + \$140.60 = \$3\ 143.66$
Total cost	$\$3\ 143.66 + \$1\ 750.75 + \$510.25 + \$625.50 = \$6\ 030.16$
Cost per casting	$\$6\ 030.16 \div 2\ 850 = \$2.12$
Profit per casting	$\$3.43 - \$2.12 =$ **$1.31** *Ans*

UNIT REVIEW

Exercise 17-3

Divide the following numbers.

1. $0.8 \div 0.2$
2. $2.4 \div 0.3$
3. $0.6 \div 0.02$
4. $4.158 \div 0.33$
5. $0.153 \div 0.05$
6. $3.794\,2 \div 1.22$
7. $0.052\,5 \div 1\frac{3}{4}$
8. $0.053\,2 \div 0.04$
9. $0.002 \div 0.250$
10. $1.512 \div 0.6$
11. $90.605\,9 \div 2.009$
12. $0.003\,36 \div 0.001\,6$
13. $94.435 \div 555.5$
14. $0.048\,75 \div \frac{13}{16}$

Divide the following numbers. Round the answers to the indicated number of decimal places.

15. $3.056\,15 \div 0.009$ (1 place)
16. $0.136 \div 0.566$ (4 places)
17. $8.508 \div 7.971$ (2 places)
18. $6\,000.8 \div 9.3$ (1 place)
19. $0.007\,2 \div 0.791$ (3 places)
20. $12.210\,04 \div \frac{7}{8}$ (5 places)

Divide the following numbers. Use the rules for dividing by a power of ten.

21. $0.76 \div 10$
22. $8.61 \div 100$
23. $0.04 \div 100$
24. $79.501 \div 1\,000$
25. $358.72 \div 10\,000$
26. $0.44 \div 0.1$
27. $0.206 \div 0.01$
28. $29.4 \div 0.001$
29. $0.002 \div 0.01$
30. $4.392\,1 \div 0.000\,01$
31. $0.075 \div 0.1$
32. $88.1 \div 1\,000$
33. $0.004 \div 0.01$
34. $0.097\,2 \div 0.000\,1$

Refer to the chart for 35–40. Find how much greater value **A** is than value **B**. Express the answer to the nearer thousandth when necessary.

	A	B
35.	8.58 kilograms ÷ 0.25	6.75 kilograms ÷ 0.375
36.	17.33 metres ÷ 1.23	15.18 metres ÷ 2.03
37.	273.56 gallons ÷ 15	291.13 gallons ÷ 37
38.	57.83 acres ÷ 7.66	42.09 acres ÷ 10.4
39.	19 hours ÷ 2.055	33 hours ÷ 5.934
40.	65.75 litres ÷ 0.56	71.02 litres ÷ 0.89

PRACTICAL APPLICATIONS

Exercise 17-4

1. Four sets of equally spaced holes are shown in the machined plate illustrated. Find dimension **A, B, C,** and **D** to the nearer thousandth centimetre.

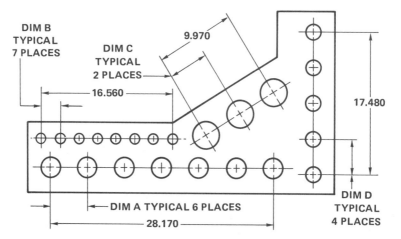

2. The amounts of trap rock delivered to a construction site during one week are listed in the table. Trap rock weighs 1.28 tons per cubic yard. Find the number of cubic yards of trap rock delivered each day. Express to the nearer hundredth.

DAY	NUMBER OF TONS DELIVERED	NUMBER OF CUBIC YARDS DELIVERED
Monday	17.25	
Tuesday	20.70	
Wednesday	18.85	
Thursday	26.40	
Friday	22.65	

3. An offset printer bases prices charged for work on the number of pages per job. The price per page is reduced as the number of pages per job is increased. Find the number of pages (one side of a sheet) printed for each of the following jobs.
   a. Job A: total printing price, $263.70, price per page, $0.045.
   b. Job B: total printing price, $368.76, price per page, $0.042.
   c. Job C: total printing price, $490.20, price per page, $0.038.

4. Often different brands of the same product are packaged in various sizes or weights. To determine which brand is the best buy, it is necessary to break the prices down to unit prices. The table lists three brands, X, Y, and Z, of the same product. Find the unit price (price per ounce) of each brand to the nearer cent.

BRAND	WEIGHT OF PACKAGE CONTENTS IN OUNCES	PRICE OF PACKAGE	UNIT PRICE PRICE PER OUNCE
X	22.5	$3.82	
Y	32.25	$4.51	
Z	25.75	$4.12	

5. A floor covering installer is contracted to install vinyl tile in a building. After measuring and finding the building floor area, the installer determines the number of tiles required for the job. Two different size tiles are to be used.

   Ten-inch square tile: each tile covers an area of 0.694 square foot. A floor area of 3 760 square feet is to be covered with 10-in square tiles.

   Fourteen-inch square tile: each tile covers an area of 1.361 square feet. A floor area of 5 150 square feet is to be covered with 14-in square tiles.

   a. Find the number of ten-inch square tiles needed to the nearer whole tile. No allowance is made for waste.
   b. Find the number of fourteen-inch square tiles needed to the nearer whole tile. No allowance is made for waste.

6. Given the following data compute the profit per casting.

   Selling price per casting, $2.64   Labor Cost, $608.50
   Number of castings, 3 100   Overhead Cost, $716.75
   Die cost, $983.25

   Material quantities and costs are listed in the table.

MATERIAL REQUIRED TO PRODUCE 3 100 CASTINGS		
MATERIAL (MOLTEN METAL)	NUMBER OF KILOGRAMS REQUIRED	COST PER KILOGRAM
Zinc	3 048 kilograms	$1.14 per kilogram
Aluminum	174.75 kilograms	$0.95 per kilogram
Copper	102.50 kilograms	$1.72 per kilogram

174 Section 3 Decimal Fractions

7. The bracket shown is part of an aircraft assembly. It is important that the bracket be as light in weight as possible. To reduce weight, equal sized holes are drilled in the bracket. The bracket weighs 5.20 kilograms before the holes are drilled. After 14 holes are drilled in the bracket, the weight is reduced by 1.26 kilograms. How many more holes of the same size must be drilled to reduce the weight of the bracket to 3.40 kilograms?

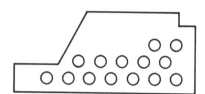

8. A home remodeler purchases hardware in the quantities listed in the table.

ITEMS	TOTAL QUANTITY PURCHASED	TOTAL COST OF PURCHASED QUANTITIES
Cabinet Hinges	12 boxes	$50.52
Drawer Pulls	15 boxes	$24.15
Cabinet Knobs	20 boxes	$36.60
Hanger Bolts	5 boxes	$14.70
Magnetic Catches	8 boxes	$18.24

The remodeler is able to reduce unit costs by quantity purchases. For a certain kitchen remodeling job, the following quantities of hardware are used:

   5 boxes of cabinet hinges

   $8\frac{1}{2}$ boxes of drawer pulls

   7 boxes of cabinet knobs

   $1\frac{1}{4}$ boxes of hanger bolts

   3 boxes of magnetic catches

Find the total hardware cost that should be charged against this job.

9. Series electrical circuits are shown. In a series circuit the total circuit resistance (R_T) equals the sum of the individual resistances.

$$R_T = R_1 + R_2 + R_3 + R_4 + R_5$$

Current in the circuit (I) equals voltage (E) applied to the circuit divided by the total resistance (R_T) of the circuit.

$$I \text{ (amperes)} = \frac{E \text{ (volts)}}{R_T \text{ (ohms)}}$$

Determine the current (amperes) in each of the circuits to 2 decimal places.

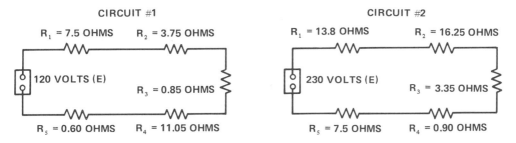

10. A welded support bar is shown. (All dimensions are in inches.)

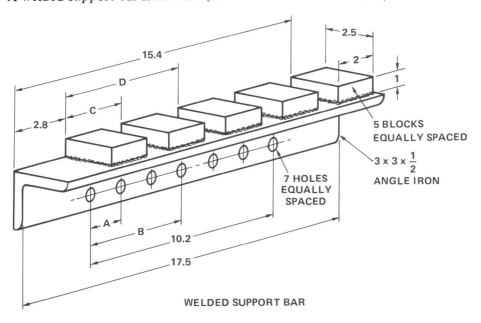

WELDED SUPPORT BAR

a. Find, in inches, dimensions **A, B, C,** and **D**.
b. The 3" x 3" x 1/2" angle iron weighs 9.4 pounds per foot of length. The 5 blocks are cut from 1" x 2" bar stock, which weighs 6.8 pounds per foot of length. The 7 holes reduce the weight a total of 0.2 pound. The weld around the blocks increases the weight a total of 0.3 pound. Find the total weight of the support bar in pounds to the nearer tenth of a pound.

UNIT 18
Powers and Roots of Decimal Fractions

OBJECTIVES

After studying this unit you should be able to

- Raise numbers to indicated powers.
- Determine whole number roots of numbers.
- Determine the square root of any positive number.
- Solve practical problems using powers and roots.
- Solve practical problems by using power and root operations in combination with one or more additional arithmetic operation.

Powers of numbers are used to find the area of square surfaces and circular sections. Volumes of cubes, cylinders, and cones are determined by applying the power operation. Determining roots of numbers is used to find the lengths of sides and heights of certain geometric figures. Both powers and roots are required operations in solving many formulas in the electrical, machine, construction, and business occupations.

MEANING OF POWERS

Two or more numbers multiplied to produce a given number are *factors* of the given number. Two factors of 8 are 2 and 4. The factors of 15 are 3 and 5. A *power* is the product of two or more equal factors. An *exponent* shows how many times a number is taken as a factor. It is written smaller than the number, above the number, and to the right of the number.

Examples:

Find the indicated powers for each of the following.

1. 4^2

   4^2 means 4 × 4. The exponent 2 shows that 4 is taken as a factor twice. It is read, "four to the second power" or "four squared."

   $4^2 = 16$ *Ans*

2. 2^5

   2^5 means 2 × 2 × 2 × 2 × 2; 2 is taken as a factor 5 times. It is read, "two to the fifth power."

   $2^5 = 32$ *Ans*

3. 0.8^3

   0.8^3 means 0.8 × 0.8 × 0.8; 0.8 is taken as a factor 3 times. It is read, "0.8 to the third power" or "0.8 cubed."

   $0.8^3 = 0.512$ *Ans*

4. 6.5^4

   6.5^4 means 6.5 × 6.5 × 6.5 × 6.5; 6.5 is taken as a factor 4 times. It is read, "6.5 to the fourth power."

   $6.5^4 = 1\ 785.062\ 5$ *Ans*

Practice

1. A carpet retailer and installer bases the price of a carpet and installation costs on the number of square yards of carpet required for a room. Find the number of square yards of carpet required for a square room whose floor plan is shown.

   $A = s^2$ where A = area

                               s = side

   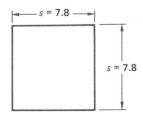

   60.84 square yards *Ans*

Section 3 Decimal Fractions

2. Find the volume (number of cubic centimetres) contained in the cube shown.

$V = s^3$ where V = volume
s = side

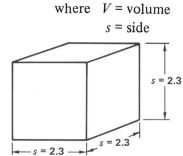

12.167 cubic centimetres *Ans*

Raise each number to the indicated power.

3. 3^2 9
4. 7^2 49
5. 5^3 125
6. 10^3 .... 1 000
7. 5.2^2 .... 27.04

Exercise 18-1

Raise the following numbers to the indicated powers.

1. 0.8^2
2. 0.8^3
3. 1^7
4. 1.3^3
5. 0.1^5
6. 2^6
7. 15.2^2
8. 9.6^3
9. 0.85^3
10. 100^3
11. 3.3^3
12. 0.2^4
13. 3^5
14. 125.25^2
15. 0.009^3

USE OF PARENTHESES

Parentheses are used as a grouping symbol. When an expression consisting of operations within parentheses is raised to a power, the operations within the parentheses are performed first. The result is then raised to the indicated power.

Example: Raise to the indicated power.

$(1.4 \times 0.3)^2$

Perform the operations within the parentheses first.

Raise to the indicated power.

$(1.4 \times 0.3)^2 = 0.42^2 = 0.176\ 4$ *Ans*

Practice

1. $(0.4 + 2.1)^3$ 15.625
2. $(6.25 - 5.75)^4$ 0.062 5
3. $(0.9 \times 5)^2$ 20.25
4. $(6.1 + 2.3)^2$ 70.56
5. $(0.77 - 0.63)^2$ 0.019 6

Parentheses which enclose a fraction indicate that both the numerator and denominator are raised to the given power.

$$\left(\frac{3}{4}\right)^2 = \frac{3^2}{4^2}$$

The same answer is obtained by dividing first and squaring second as by squaring both terms first and dividing second.

Exercise 18-2

Raise the following expressions to the indicated powers.

1. $\left(\frac{1}{2}\right)^2$
2. $\left(\frac{1}{2}\right)^3$
3. $(10 \times 1.6)^3$
4. $(0.74 + 1.26)^5$
5. $(33.54 - 21.27)^2$
6. $\left(\frac{3}{8}\right)^2$
7. $\left(\frac{3}{5}\right)^3$
8. $(14.8 - 11.8)^5$
9. $(9.9 \times 0.01)^2$
10. $(2 + 0.5)^2$
11. $(2 + 0.5)^4$
12. $\left(\frac{13}{20}\right)^2$
13. $(187.5 - 186)^3$
14. $(1\ 000 \times 0.001)^8$
15. $(0.36 \times 18.14)^2$
16. $306 \times 0.02)^2$
17. $\left(\frac{9}{10}\right)^3$
18. $(14.3 - 14.1)^4$

COMBINING POWER OPERATIONS WITH OTHER ARITHMETIC OPERATIONS

Many actual on-the-job applications require a combination of power operations with other operations. It is essential that the solution to the complete problem be determined before computations are made.

Example: A construction technician finds the weight of 8 concrete pier footings. Footings distribute the load (weight) of walls and support columns over a larger area. The concrete used for the pier footings weighs 2 380 kilograms per cubic metre.

$$V = s^3 \quad \text{where} \quad V = \text{volume}$$
$$s = \text{side}$$

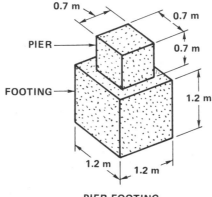

PIER FOOTING

Volume of pier: 0.7 m × 0.7 m × 0.7 m = 0.343 m³

Volume of footing: 1.2 m × 1.2 m × 1.2 m = 1.728 m³

Total volume: 0.343 m³ + 1.728 m³ = 2.071 m³

Weight of one pier footing: $\dfrac{2.071 \text{ m}^3}{1} \times \dfrac{2\,380 \text{ kg}}{\text{m}^3} = 4\,928.98 \text{ kg}$

Weight of eight pier footings: 4 928.98 kg × 8 = **39 431.84 kg** *Ans*

DESCRIPTION OF ROOTS

The *root* of a number is a quantity which is taken two or more times as an equal factor of the number. Finding a root is the opposite operation of finding a power.

The *radical symbol* ($\sqrt{}$) is used to indicate a root of a number. The *index* indicates the amount of times that a root is to be taken as an equal factor to produce the given number. The index is written smaller than the number, above and to the left of the radical symbol. The index 2 is usually omitted. For example $\sqrt{9}$ means to find the number which can be multiplied by itself and equal 9. In the expression $\sqrt[3]{8}$, the index, 3, indicates the root is taken as a factor 3 times to equal 8. The cube root of 8 is 2.

Unit 18 Powers and Roots of Decimal Fractions 181

Examples:

Find the indicated roots.

1. $\sqrt{36}$ ⟶ $6 \times 6 = 36$; therefore, $\sqrt{36} = 6$ Ans
2. $\sqrt{144}$ ⟶ $12 \times 12 = 144$; therefore, $\sqrt{144} = 12$ Ans
3. $\sqrt[3]{8}$ ⟶ $2 \times 2 \times 2 = 8$; therefore, $\sqrt[3]{8} = 2$ Ans
4. $\sqrt[3]{125}$ ⟶ $5 \times 5 \times 5 = 125$; therefore, $\sqrt[3]{125} = 5$ Ans
5. $\sqrt[4]{81}$ ⟶ $3 \times 3 \times 3 \times 3 = 81$; therefore, $\sqrt[4]{81} = 3$ Ans

Practice

1. A lumber shed in the shape of a cube is to be constructed so that it contains a volume of 1 000 cubic feet. Find the length of one side.

$s = \sqrt[3]{V}$ where s = side
             V = volume

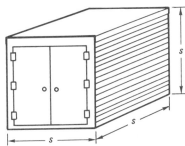

10 ft Ans

Find the indicated roots.

2. $\sqrt{169}$ 13
3. $\sqrt[8]{1}$ 1
4. $\sqrt[4]{256}$ 4
5. $\sqrt[3]{1\,000\,000}$ 100
6. $\sqrt[4]{625}$ 5

Exercise 18-3

Determine the indicated roots of the following numbers.

1. $\sqrt{25}$ 3. $\sqrt{49}$ 5. $\sqrt[3]{27}$
2. $\sqrt{100}$ 4. $\sqrt[3]{1}$ 6. $\sqrt[4]{16}$

7. $\sqrt{144}$
8. $\sqrt{256}$
9. $\sqrt[3]{64}$
10. $\sqrt[4]{10\,000}$
11. $\sqrt{196}$
12. $\sqrt[5]{32}$
13. $\sqrt{81}$
14. $\sqrt[6]{64}$
15. $\sqrt{10\,000}$

EXPRESSIONS ENCLOSED WITHIN THE RADICAL SYMBOL

The radical symbol is a grouping symbol. An expression consisting of operations within the radical symbol is done using the order of operations. The operations within the radical symbol are performed first. Next find the root.

Examples:
1. $\sqrt{3 \times 12} \longrightarrow \sqrt{36} = \sqrt{6 \times 6}$; therefore, $\sqrt{3 \times 12} = 6$ *Ans*
2. $\sqrt[3]{9.2 + 54.8} \longrightarrow \sqrt[3]{64} = \sqrt[3]{4 \times 4 \times 4}$; therefore, $\sqrt[3]{9.2 + 54.8} = 4$ *Ans*
3. $\sqrt{148.2 - 27.2} \longrightarrow \sqrt{121} = \sqrt{11 \times 11}$; therefore, $\sqrt{148.2 - 27.2} = 11$ *Ans*

A radical symbol which encloses a fraction indicates that the roots of both the numerator and denominator are to be taken.

$$\sqrt{\frac{900}{4}} = \frac{\sqrt{900}}{\sqrt{4}}$$

The same answer is obtained by dividing first and extracting the root second, as by extracting both roots first and dividing second.

Example: An electrician determines current by the following formula:
$$I = \sqrt{\frac{P}{R}}$$

Find the current (I) in amperes of an appliance which has a resistance (R) of 4 ohms and consumes 900 watts of power (P).

$$I = \sqrt{\frac{P}{R}}$$
$$I = \sqrt{\frac{900}{4}}$$
$$I = \sqrt{225}$$
$$I = 15 \text{ amperes } Ans$$

Practice

1. $\sqrt{632 - 551}$ 9
2. $\sqrt[4]{1.125 \times 72}$ 3
3. $\sqrt[3]{\dfrac{2.56}{0.04}}$ 4

Exercise 18-4

Determine the indicated roots of the following expressions.

1. $\sqrt{5 \times 20}$
2. $\sqrt{3.1 + 12.9}$
3. $\sqrt{10.7 - 6.7}$
4. $\sqrt{\dfrac{50}{2}}$
5. $\sqrt{0.7 \times 70}$
6. $\sqrt[3]{3 \times 9}$
7. $\sqrt[3]{40 \times 0.2}$
8. $\sqrt{33.55 + 2.45}$
9. $\sqrt{87.64 + 12.36}$
10. $\sqrt{\dfrac{360}{2.5}}$
11. $\sqrt{\dfrac{23}{23}}$
12. $\sqrt[3]{127.3 - 63.3}$
13. $\sqrt[3]{12.5 \times 10}$
14. $\sqrt{320 \times 0.45}$
15. $\sqrt[4]{\dfrac{54.4}{3.4}}$
16. $\sqrt{99.03 + 125.97}$
17. $\sqrt[3]{101.7 + 23.3}$
18. $\sqrt[5]{6.25 \times 0.16}$

GENERAL METHOD OF COMPUTING SQUARE ROOTS

The square root examples shown have all consisted of perfect squares. *Perfect squares* are numbers which have whole number square roots. These roots are relatively easy to determine by observation.

Most numbers do not have whole number square roots; therefore, a procedure must be used in computing square roots of these numbers.

Section 3 Decimal Fractions

Example: Find the square root of 5 410.218 to 1 decimal place.

Beginning at a decimal point, group the digits in pairs.

Annex a zero to the 8 in order to form a pair of digits.

Place the decimal point directly above the decimal point of the number.

$$\sqrt{54\ 10\ .\ 21\ 80}$$

Find the largest perfect square that can be subtracted from the first digit or pair of digits. The largest perfect square is 49. Write the square root of this perfect square above the first digit or group of digits. The square root of 49 is 7.

$$\begin{array}{r} 7\ \ \ \ .\ \ \ \ \ \ \ \ \ \\ \sqrt{54\ 10\ .\ 21\ 80} \\ \underline{49\ \ \ \ \ \ \ \ \ \ \ \ \ } \\ 5\ \ \ \ \ \ \ \ \ \ \ \ \end{array}$$

Bring down the next pair of digits (10) and place by the remainder (5).

$$\begin{array}{r} 7\ \ \ \ .\ \ \ \ \ \ \ \ \ \\ \sqrt{54\ 10\ .\ 21\ 80} \\ \underline{49\ \ \ \ \ \ \ \ \ \ \ \ \ } \\ 5\ 10\ \ \ \ \ \ \ \ \end{array}$$

Double the partial root and use this number as a trial divisor.

$$2 \times 7 = 14$$

Divide the remainder by this trial divisor, disregarding the last digit of the remainder.

$$51 \div 14 = 3$$

$$\begin{array}{r} 7\ \ \ \ .\ \ \ \ \ \ \ \ \ \\ \sqrt{54\ 10\ .\ 21\ 80} \\ 49\ \ \ \ \ \ \ \ \ \ \ \ \ \\ 14\ \overline{)\ 5\ 10\ \ \ \ \ \ \ } \end{array}$$

Annex the quotient as the next figure in the square root. Also annex the same digit to the trial divisor.

Annex 3 to the root and to the trial divisor.

Multiply the complete divisor by the digit which was annexed to the root, and subtract.

Multiply: 3 × 143 = 429
Subtract: 510 − 429 = 81

Repeat the process until the desired number of decimal places is obtained.

$$\begin{array}{r} 7\ 3\ .\ \ \ \ \ \ \ \\ \sqrt{54\ 10\ .\ 21\ 80} \\ 49\ \ \ \ \ \ \ \ \ \ \ \ \ \\ 143\ \overline{)\ 5\ 10\ \ \ \ \ \ \ } \end{array}$$

$$\begin{array}{r} 7\ 3\ .\ \ \ \ \ \ \ \\ \sqrt{54\ 10\ .\ 21\ 80} \\ 49\ \ \ \ \ \ \ \ \ \ \ \ \ \\ 143\ \overline{)\ 5\ 10\ \ \ \ \ \ \ } \\ \underline{4\ 29\ \ \ \ \ \ \ } \\ 81\ \ \ \ \ \ \ \end{array}$$

Bring down 21.

Double 73; 2 × 73 = 146

Divide 812 by 146; 812 ÷ 146 = 5.

Annex 5 to the root and to the trial divisor.

Multiply: 5 × 1 465 = 7 325.

Subtract: 8 121 − 7 325 = 796.

Bring down 80.

Double 735; 2 × 735 = 1 470.

Divide 7 968 by 1 470; 7 968 ÷ 1 470 = 5.

Annex 5 to the root and to the trial divisor.

Multiply: 5 × 14 705 = 73 525.

Subtract: 79 680 − 73 525 = 6 155.

Round 73.55 to 1 decimal place,

```
              7  3 . 5  5
          √ 54 10 . 21 80
            49
      143 ) 5 10
            4 29
    1 465 ) 81 21
            73 25
   14 705 ) 7 96 80
            7 35 25
              61 55
```

73.6 *Ans*

Practice 1: Determine the square root of 923.7 to 2 decimal places.

```
            3  0 . 3  9  2
        √  9 23 . 70 00 00
           9
      60 ) 0 23
            00
     603 ) 23 70
           18 09
   6 069 ) 5 61 00
           5 46 21
  60 782 ) 14 79 00
           12 15 64
            2 63 36
```

Round 30.392 to 2 decimal places,

30.39 *Ans*

Practice 2: Determine the square root of 0.003 9 to 3 decimal places.

```
              0 . 0  6  2  4
        √  0 . 00 39 00 00
                  36
        122 )     3 00
                  2 44
      1 244 )     56 00
                  49 76
                   6 24
```

Round 0.062 4 to 3 decimal places,

0.062 *Ans*

Exercise 18-5

Determine the square roots of the following numbers to the indicated number of decimal places.

1. 123 (2 places)
2. 3.182 (3 places)
3. 0.654 1 (4 places)
4. 0.08 (2 places)
5. 987.6 (1 place)
6. 427.52 (2 places)
7. 5.060 8 (3 places)
8. 2.581 (3 places)
9. 0.005 6 (4 places)
10. 6 085 (0 places)
11. 3 258.83 (2 places)
12. 0.000 15 (2 places)
13. 73.808 (3 places)
14. 9 604.1 (0 places)
15. 9 216.47 (2 places)
16. 7.035 89 (4 places)
17. 69 281 (1 place)
18. 0.024 7 (3 places)

COMBINING ROOT OPERATIONS WITH OTHER ARITHMETIC OPERATIONS

In occupational uses, root operations are most often applied in combination with other operations. Before making any computations, think a problem through to determine the steps necessary in its solution.

> **Example:** A landscaper is contracted to plant shrubs along two edges of a square plot of land. The plot contains an area of 19 600 square feet. Shrubs are planted at each of the four corners. Then shrubs are spaced 3.5 feet apart along two edges as shown. Find the total number of shrubs needed.
>
> $s = \sqrt{A}$ where s = side
> A = area
>
>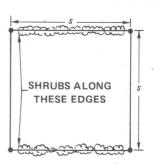
>
> SHRUBS ALONG THESE EDGES
>
> | | |
> |---|---|
> | Length of one side | $\sqrt{19\ 600}$ sq ft = 140 ft |
> | Number of shrubs along one side (There is one more shrub than there are spaces.) | 140 ft ÷ 3.5 ft = 40 spaces
40 + 1 = 41 shrubs |
> | Total for 2 sides | 41 shrubs × 2 = 82 shrubs *Ans* |

Practice

A welder finds the material needed to fabricate the steel storage tank shown. The specifications call for the tank to be a cube capable of holding 935 gallons of oil. One cubic foot contains 7.48 gallons. Find the length of one side of the tank.

$$s = \sqrt[3]{V} \quad \text{where} \quad \begin{array}{l} s = \text{side} \\ V = \text{volume} \end{array}$$

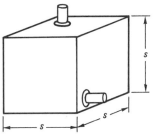

5 feet *Ans*

UNIT REVIEW

Exercise 18-6

Raise the following numbers to the indicated powers.

1. 6^2
2. 2^3
3. 20^2
4. 3.7^2
5. 0.9^2
6. 3.1^3
7. 16.6^2
8. 8.5^3
9. 0.61^3
10. 100^3
11. 2.2^3
12. 0.3^4
13. 2^5
14. 113.06^2
15. $0.000\ 8^2$

Raise the following expressions to the indicated powers.

16. $(0.6 \times 7)^2$
17. $(3.8 + 6.4)^2$
18. $(0.36 - 0.11)^2$
19. $\left(\dfrac{2}{5}\right)^2$
20. $\left(\dfrac{2}{5}\right)^3$
21. $(0.88 + 2.09)^2$
22. $(0.39 + 0.61)^6$
23. $(6.4 \times 0.02)^2$
24. $(15.2 - 9.2)^4$
25. $(200 \times 0.01)^5$
26. $\left(\dfrac{7}{100}\right)^2$
27. $(19.9 - 17.6)^3$
28. $(0.53 + 6.47)^3$
29. $(5\ 000 \times 0.000\ 2)^6$

Determine the whole number roots of the following numbers as indicated.

30. $\sqrt{16}$
31. $\sqrt{64}$
32. $\sqrt{81}$
33. $\sqrt[3]{8}$
34. $\sqrt[3]{125}$
35. $\sqrt[6]{1}$
36. $\sqrt{121}$
37. $\sqrt[4]{81}$
38. $\sqrt{196}$

188 Section 3 Decimal Fractions

39. $\sqrt{225}$
40. $\sqrt[4]{10\,000}$
41. $\sqrt[3]{27}$
42. $\sqrt[3]{216}$
43. $\sqrt[6]{64}$
44. $\sqrt[3]{343}$

Determine the whole number roots of the following expressions as indicated.

45. $\sqrt{3 \times 12}$
46. $\sqrt{7.03 + 1.97}$
47. $\sqrt{18.8 - 2.8}$
48. $\sqrt{\dfrac{24}{6}}$
49. $\sqrt{0.8 \times 80}$
50. $\sqrt{12.5 \times 2}$
51. $\sqrt[3]{92.77 + 32.23}$
52. $\sqrt{221.3 - 77.3}$
53. $\sqrt{0.45 \times 20}$
54. $\sqrt{\dfrac{615.6}{7.6}}$
55. $\sqrt[3]{\dfrac{1.08}{0.04}}$
56. $\sqrt[4]{10.125 \times 8}$
57. $\sqrt{19.09 + 101.91}$
58. $\sqrt[5]{\dfrac{8}{0.25}}$

Determine the square roots of the following numbers to the indicated number of decimal places.

59. 247 (2 places)
60. 5.319 (3 places)
61. 0.821 4 (4 places)
62. 706.1 (1 place)
63. 9.623 4 (3 places)
64. 0.006 2 (4 places)
65. 6 599 (2 places)
66. 0.000 2 (2 places)
67. 87.705 (3 places)
68. 64 404 (1 place)
69. 0.031 8 (3 places)
70. 26.204 (2 places)
71. 1.101 01 (4 places)
72. 6 304.04 (2 places)

Refer to the chart for 73–78. Find how much greater value **A** is than value **B**. Express the answers to 2 decimal places.

	A	B
73.	(5.6 inches)²	(4.8 inches)²
74.	(0.82 kilometres)²	(0.51 kilometres)²
75.	$\sqrt{14.42}$ sq yd	$\sqrt{8.11}$ sq yd
76.	(9.3 metres)³	(0.8 metres)³
77.	$\sqrt{270.81}$ sq in	$\sqrt{94.08}$ sq in
78.	(1.73 yards)³	(0.32 yards)³

PRACTICAL APPLICATIONS

Exercise 18-7

1. In the table shown, the lengths of the sides of squares are given. Find the areas of the squares to two decimal places.

	LENGTH OF SIDES (s)	AREA (A)
a.	1.25 ft	
b.	32.3 cm	
c.	7.3 yd	
d.	0.66 km	
e.	210.2 ft	

   $A = s^2$ where A = area, s = side

   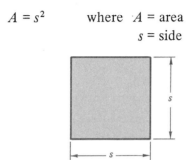

2. In the table shown, the lengths of the sides of cubes are given. Determine the volumes of the cubes to three decimal places.

	LENGTH OF SIDES (s)	VOLUME (V)
a.	9.7 mm	
b.	14.01 in	
c.	6.6 ft	
d.	10.75 cm	
e.	0.12 ft	

   $V = s^3$ where V = volume, s = side

   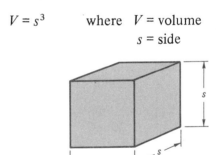

3. In the table shown, the areas of squares are given. Determine the lengths of the sides of the squares to two decimal places where necessary.

	AREA (A)	LENGTH OF SIDES (s)
a.	125 sq ft	
b.	9 km²	
c.	57.75 sq in	
d.	0.88 m²	
e.	247 sq ft	

   $s = \sqrt{A}$ where s = side, A = area

   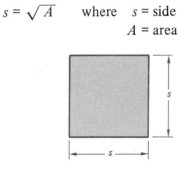

Section 3 Decimal Fractions

4. In the table shown, the volumes of cubes are given. Determine the lengths of the sides of the cubes.

	VOLUME (V)	LENGTH OF SIDES (s)
a.	27 cm³	
b.	125 cu ft	
c.	1 m³	
d.	8 cu yd	
e.	216 cu ft	

$s = \sqrt[3]{V}$ where s = side
V = volume

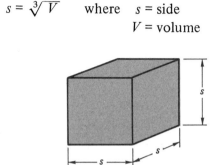

5. Find the current in amperes of the circuits listed in the table. Express the answer to the nearer tenth ampere when necessary.

$$I = \sqrt{\frac{P}{R}}$$
where I = current in amperes
P = power in watts
R = resistance in ohms

CIRCUIT	POWER (P)	RESISTANCE (R)	CURRENT (I)
a	288 watts	2 ohms	
b	2 320 watts	5.8 ohms	
c	3 050 watts	9.6 ohms	
d	5 240 watts	14.7 ohms	

Note: Use these formulas for 6–12.

$A = s^2$ where A = area
$V = s^3$ V = volume
$s = \sqrt{A}$ s = side
$s = \sqrt[3]{V}$

6. Compute the cost of filling a hole 5.5 metres long, 5.5 metres wide, and 5.5 metres deep. The cost of fill soil is $4.75 per cubic metre.

7. A paving contractor, in determining the cost of a job, finds the number of square feet (area) to be paved. The shaded area which surrounds a building shown is paved. Find the number of square feet of pavement required.

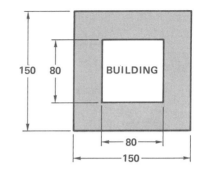

8. A plot of land consists of two parcels, A and B. Both parcels are squares. The plot has a total area of 22 500 square metres. Find, in metres, length C.

$$s = \sqrt{A}$$ where s = side
A = area

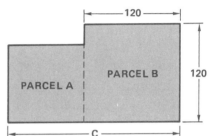

9. A steel storage tank is to be fabricated by a welder. The tank is to be made in the shape of a cube capable of holding 2 565.64 gallons of fuel. One cubic foot contains 7.48 gallons. Find the length of one side of the storage tank.

$$s = \sqrt[3]{V}$$ where s = side
V = volume

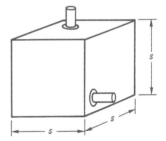

10. A plate is designed to contain 210 square centimetres of metal after the circular cutout is removed. A designer finds the length of the radius of the cutout to determine the size of the circle to be removed. Find, in centimetres, the length of the required radius to 2 decimal places. Note: A radius is a straight line that connects the center of a circle with a point on the circle.

$$R = \sqrt{\frac{A}{3.1416}}$$ where R = radius
A = area of circle

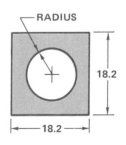

11. A fabric sample maker lays out and punches eyelet holes in the square piece of canvas. The holes are punched around all four edges of the canvas, 2 inches in from each edge. All holes are equally spaced at 2.5 inches apart. The total area of the canvas is 3.16 square yards (4 096 square inches). Determine the total number of eyelet holes required. Note: Do not count the number of holes shown in the drawing; the actual number is not shown.

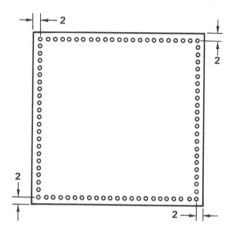

12. A materials estimator finds the weight of aluminum needed for the casting shown. Aluminum weighs 0.097 5 pound per cubic inch. Find, to the nearer pound, the weight of aluminum required for 15 castings. All measurements are in inches.

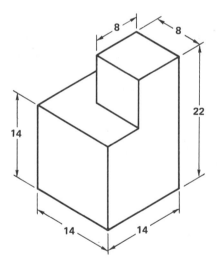

UNIT 19
Combined Operations with Decimal Fractions

OBJECTIVES

After studying this unit you should be able to

- Write decimal or fraction equivalents using a decimal equivalent table.
- Determine nearer fraction equivalent using a decimal equivalent table.
- Solve arithmetic expressions using the proper order of operations.
- Solve practical applied problem formulas by applying the proper order of operations.

TABLE OF DECIMAL EQUIVALENTS

Generally, fractional machine, mechanical, and sheet metal blueprint dimensions are given in multiples of 64ths of an inch. Carpenters, cabinetmakers, and many other woodworkers measure in multiples of 32nds of an inch.

In certain occupations, it is often necessary to express fractional dimensions as decimal dimensions. A machinist is required to express fractional dimensions as decimal equivalents for machine settings in making a part. Decimal dimensions are expressed as fractional dimensions if fractional measuring devices are used. A patternmaker may express decimal dimensions to the nearer equivalent 64th inch.

Using a decimal equivalent table saves time and reduces the chance of error. Decimal equivalent tables are widely used in business and industry. They are posted as large wall charts in work areas and are available as pocket size cards. Skilled workers memorize many of the equivalents after using decimal equivalent tables.

The decimals listed in the table are given to six places. In actual practice, a decimal is rounded to the degree of precision desired for a particular application.

DECIMAL EQUIVALENT TABLE

Fraction	Decimal	Fraction	Decimal
1/64	0.015 625	33/64	0.515 625
1/32	0.031 25	17/32	0.531 25
3/64	0.046 875	35/64	0.546 875
1/16	0.062 5	9/16	0.562 5
5/64	0.078 125	37/64	0.578 125
3/32	0.093 75	19/32	0.593 75
7/64	0.109 375	39/64	0.609 375
1/8	0.125	5/8	0.625
9/64	0.140 625	41/64	0.640 625
5/32	0.156 25	21/32	0.656 25
11/64	0.171 875	43/64	0.671 875
3/16	0.187 5	11/16	0.687 5
13/64	0.203 125	45/64	0.703 125
7/32	0.218 75	23/32	0.718 75
15/64	0.234 375	47/64	0.734 375
1/4	0.25	3/4	0.75
17/64	0.265 625	49/64	0.765 625
9/32	0.281 25	25/32	0.781 25
19/64	0.296 875	51/64	0.796 875
5/16	0.312 5	13/16	0.812 5
21/64	0.328 125	53/64	0.828 125
11/32	0.343 75	27/32	0.843 75
23/64	0.359 375	55/64	0.859 375
3/8	0.375	7/8	0.875
25/64	0.390 625	57/64	0.890 625
13/32	0.406 25	29/32	0.906 25
27/64	0.421 875	59/64	0.921 875
7/16	0.437 5	15/16	0.937 5
29/64	0.453 125	61/64	0.953 125
15/32	0.468 75	31/32	0.968 75
31/64	0.484 375	63/64	0.984 375
1/2	0.5	1	1.

Unit 19 Combined Operations With Decimal Fractions 195

Example: Find the nearer fraction equivalents of the decimal dimensions given on the drawing of the wood pattern shown.

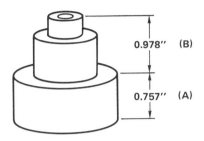

Dimension A is between 0.750" and 0.765 625". Subtract to find the closer dimension.

0.765 625" − 0.757" = 0.008 625"
0.757" − 0.750" = 0.007"

Dimension A is closer to 0.750". Find the fraction equivalent for 0.750".

$0.757'' \approx \frac{3''}{4}$ Ans

Dimension B is between 0.968 75" and 0.984 375". Subtract to find the closer dimension.

0.984 375" − 0.978" = 0.006 375"
0.978" − 0.968 75" = 0.009 25"

Dimension B is closer to 0.984 375". Find the fraction equivalent for 0.984 375".

$0.978'' \approx \frac{63''}{64}$ Ans

Practice

1. Find the decimal equivalent of $\frac{11}{16}$. 0.687 5

2. Find the decimal equivalent of $\frac{19}{32}$. 0.593 75

3. Determine the nearer fraction equivalent of 0.289 .... $\frac{9}{32}$

4. Determine the nearer fraction equivalent of 0.555 .... $\frac{9}{16}$

Exercise 19-1

Find the decimal or fraction equivalents of the following number using the decimal equivalent table.

1. $\dfrac{9}{32}$
2. $\dfrac{15}{16}$
3. $\dfrac{11}{32}$
4. $\dfrac{7}{16}$
5. $\dfrac{5}{8}$
6. $\dfrac{5}{64}$
7. $\dfrac{3}{8}$
8. $\dfrac{43}{64}$
9. 0.281 25
10. 0.546 875
11. 0.781 25
12. 0.437 5
13. 0.093 75
14. 0.687 5
15. 0.968 75
16. 0.390 625

Determine the nearer fraction equivalents of the following decimals using the decimal equivalent table.

17. 0.209
18. 0.498
19. 0.068
20. 0.601
21. 0.351
22. 0.971
23. 0.805
24. 0.053
25. 0.727
26. 0.243
27. 0.600
28. 0.992
29. 0.416
30. 0.459
31. 0.148

32. The profile gauge shown is dimensioned fractionally in inches. Use the table of decimal equivalents. Express dimensions **A–I** in decimal form. Round the answers to 3 decimal places where necessary.

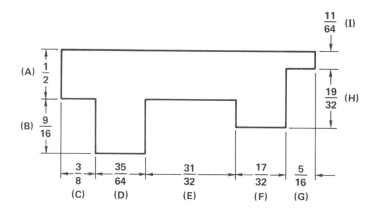

33. The hole locations in the bracket shown are dimensioned decimally in inches. Use the table of decimal equivalents. Express dimensions **A–H** in fractional form. Round the answers to the nearer 1/64 inch.

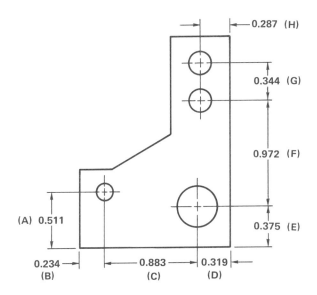

COMBINED OPERATIONS OF DECIMAL FRACTIONS

Combined operation problems are given as arithmetic expressions in this unit. Practical applications problems are based on formulas found in various occupational textbooks, handbooks, manuals, and other related reference materials.

The proper order of operations including powers and roots must be understood to solve expressions made up of any combination of the six arithmetic operations. Study the following order of operations.

1. Do all operations within the grouping symbol first. Parentheses, the fraction bar and the radical symbol are used to group numbers. If an expression contains parentheses within parentheses or brackets do the work within the innermost parentheses first.
2. Do powers and roots next. The operations are performed in the order in which they occur. If a root consists of two or more operations within the radical symbol, perform all the operations within the radical symbol, then extract the root.
3. Do multiplication and division next in the order in which they occur.
4. Do addition and subtraction last in the order in which they occur.

Examples:

1. Find the value of 8.14 + 3.6 × 0.8 − 1.37

Multiply.	8.14 + 3.6 × 0.8 − 1.37
Add.	8.14 + 2.88 − 1.37
Subtract.	11.02 − 1.37
	9.65 *Ans*

2. Find the value of (18.36 − 5.93) ÷ (12 + 6.8 ÷ 2.5) to 2 decimal places.

   Grouping Symbol operations are done first.

   $$(18.36 - 5.93) \div (12 + 6.8 \div 2.5)$$
   $$12.43 \div 14.72$$
   $$\mathbf{0.84} \; Ans$$

3. Find the value of $9.6 + \dfrac{18.54 + (12 \times 0.4)^2}{68 \times 0.08 - \sqrt{2.25}}$ to 3 decimal places.

   Grouping Symbol operations are done first. $9.6 + \dfrac{18.54 + (12 \times 0.4)^2}{68 \times 0.08 - \sqrt{2.25}}$

   a. Perform the work in [18.54 + (12 × 0.4)²]
      1. Multiply: (12 × 0.4) = 4.8
      2. Square: 4.8² = 23.04
      3. Add: 18.54 + 23.04 = 41.58

   b. Perform the work in (68 × 0.08 − $\sqrt{2.25}$)
      1. Extract the square root: $\sqrt{2.25}$ = 1.5
      2. Multiply: 68 × 0.08 = 5.44
      3. Subtract: 5.44 − 1.5 = 3.94

Divide.	9.6 + 41.58 ÷ 3.94
Add.	9.6 + 10.553
	20.153 *Ans*

Practice

1. A series-parallel electrical circuit is shown. The complete circuit consists of 2 minor circuits each connected in parallel. The 2 minor circuits are then connected in series. Use the following formula to compute the total resistance (R_T) of this circuit. Express the answer to the nearer tenth ohm.

$$R_T = \frac{1}{\frac{1}{R_1} + \frac{1}{R_2} + \frac{1}{R_3}} + \frac{1}{\frac{1}{R_4} + \frac{1}{R_5} + \frac{1}{R_6} + \frac{1}{R_7}}$$

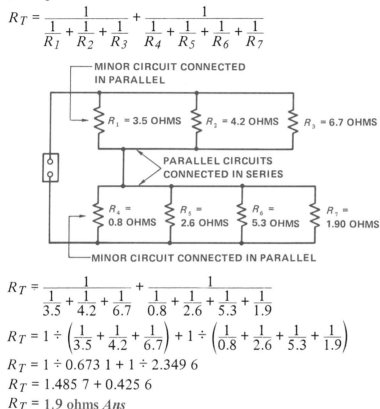

$$R_T = \frac{1}{\frac{1}{3.5} + \frac{1}{4.2} + \frac{1}{6.7}} + \frac{1}{\frac{1}{0.8} + \frac{1}{2.6} + \frac{1}{5.3} + \frac{1}{1.9}}$$

$$R_T = 1 \div \left(\frac{1}{3.5} + \frac{1}{4.2} + \frac{1}{6.7}\right) + 1 \div \left(\frac{1}{0.8} + \frac{1}{2.6} + \frac{1}{5.3} + \frac{1}{1.9}\right)$$

$R_T = 1 \div 0.673\ 1 + 1 \div 2.349\ 6$

$R_T = 1.485\ 7 + 0.425\ 6$

$R_T = 1.9$ ohms *Ans*

2. A surveyor wishes to determine the distance (AB) between two corners (points A and B) of a lot as shown. A building between the two corners prevents the taking of a direct measurement. The surveyor makes measurements and locates a stake at point C where distance AC is perpendicular to distance BC. Perpendicular means that AC and BC meet at a 90° angle. Distance AC is measured as 105.5 feet and distance BC is measured as 118 feet. Find, to the nearer tenth foot, distance AB.

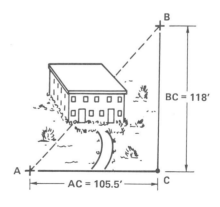

$AB = \sqrt{(AC)^2 + (BC)^2}$

158.3 feet *Ans*

200 Section 3 Decimal Fractions

3. A flat is ground on the hardened steel pin shown. A tool and die-maker computes the depth of material to be ground from the top of the pin to produce the required length of the flat.

$$C = \frac{D}{2} - 0.5 \times \sqrt{4 \times \left(\frac{D}{2}\right)^2 - F^2}$$

where C = depth of material to be removed (depth of cut)
D = diameter
F = length of flat

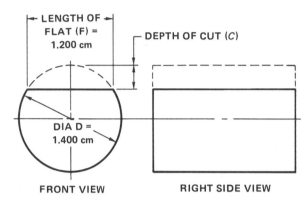

Find the depth of material removed (depth of cut) to 3 decimal places.

0.339 cm *Ans*

Exercise 19-2

Solve the following combined operations expressions. Round the answers to 2 decimal places.

1. $0.187 + 16.3 \times 1.02$
2. $\frac{4.23}{6} - 0.98 \times 0.3$
3. $20 \times 0.86 - 80.4 \div 6$
4. $24.78 + 9.07 \times 0.5$
5. $(24.78 + 9.07) \times 0.5$
6. $(18.8 - 13.3) \times (2.7 + 9.1)$
7. $40.87 + 16.04 - 3.3^2 \div 6$
8. $40.87 + (16.04 - 3.3^2) \div 6$
9. $(0.73 - 0.37)^2 \times 10.4$
10. $28.39 + (50.6 \div 12 \times 0.8 + 6)^2$
11. $0.051 + 2 \times \sqrt{25} - 6.062$
12. $\left(\frac{21.3}{7.1}\right)^3 + 14.4 + 2.2^2$
13. $\frac{21.3}{7.2^3} + (14.4 + 2.2)^2$
14. $22.76 \div \sqrt{12.32 + 1.76}$
15. $(4.31 \times 0.6)^2 \div (5.96 - 1.05)$
16. $(\sqrt{0.65} + 0.78 \times 2.4)^2$
17. $(2.39 \times 0.9)^2 \div (1.05 - 0.83)$
18. $2.39 \times (0.9 \div 1.05)^2 - 0.83$
19. $0.875 + 0.225 \times \frac{50}{\sqrt{16}}$
20. $0.25 \times \left(\frac{\sqrt{49} - 2.4}{3.8}\right) + 0.99$
21. $0.25 \times \frac{\sqrt{49} - 2.4 + 0.99}{3.8}$
22. $\frac{\sqrt{80.9} \times 3.7}{16.4 \times 1.35} + \left(\frac{18.8}{4.7}\right)^2$
23. $23.67 - \sqrt{\frac{8.63 \times 5.1}{6.5^2 - 0.59}} \times 0.9$
24. $16.79 + \frac{(32.6 \times 0.3)^2}{\sqrt{4.3} + \sqrt{8}} - 2.1$

UNIT REVIEW

Exercise 19-3

Find the decimal or fraction equivalents of the following numbers using the decimal equivalent table.

1. $\dfrac{3}{32}$
2. $\dfrac{9}{16}$
3. $\dfrac{19}{32}$
4. $\dfrac{23}{64}$
5. 0.812 5
6. 0.296 875
7. 0.609 375
8. 0.437 5
9. 0.703 125
10. 0.156 25
11. 0.468 75
12. 0.578 125

Determine the nearer fraction equivalents of the following decimals using the decimal equivalent table.

13. 0.070
14. 0.522
15. 0.603
16. 0.239
17. 0.877
18. 0.519
19. 0.205
20. 0.711
21. 0.834
22. 0.966
23. 0.484
24. 0.099
25. 0.025
26. 0.801
27. 0.658

Solve the following combined operations expressions. Round the answers to 2 decimal places.

28. $12.08 - 8.74 \times 0.6$
29. $0.98 \times 13 - 14 \div 2.2$
30. $9.34 - 0.7 \times \dfrac{8.08}{15.2}$
31. $1.16 \times (37.81 - 11.02 \times 0.6)$
32. $6.88 + (23.23 - 4.2^2) \div 0.8$
33. $(0.3 - 0.06)^2 \times 12.3$
34. $19.5 - (100 \div 12.5 \times 0.3)^2$
35. $0.123 + 3 \times \sqrt{49} - 1.015$
36. $16.06 \div \sqrt{0.98 + 1.14}$
37. $\left(\dfrac{84.4}{21.1}\right)^3 + 16 \div 2.5$
38. $(\sqrt{4.7 + 0.12} + 0.64)^2$
39. $0.912 - 0.098 \times \dfrac{\sqrt{81}}{7}$
40. $(6.93 \times 0.5)^2 \div (87.5 - 63.2)$
41. $0.75 \times \left(\dfrac{\sqrt{36} - 3.1}{1.7}\right) + 8.34$
42. $14.33 + \dfrac{(13.1 \times 0.9)^2}{\sqrt{2.6} - 0.07} - 0.88$
43. $105.6 - \sqrt{\dfrac{3.07 \times 6}{2.25 - 0.88^2}} \times 1.7$

PRACTICAL APPLICATIONS

Exercise 19-4

1. A painting contractor measures the width (W), length (L), and height (H) of a building to be painted. The number of square feet covered by one gallon of paint (C) is estimated to be 550 square feet per gallon. Determine the number of gallons of paint required for the job to the nearer full gallon. All dimensions are in feet.

$$N = \frac{2 \times (W \times H + L \times H)}{C}$$

where N = number of gallons
L = length
W = width
H = height
C = square feet covered by one gallon

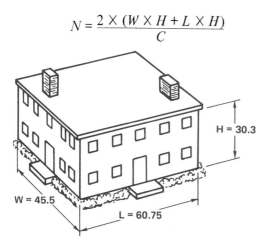

2. The bookkeeper for a small trucking firm finds the yearly depreciation of company vehicles. The bookkeeper uses the appraisal method of depreciation. Under this method, yearly depreciation is based on the fraction of the life of the vehicle used in one year. The following formula is used to compute yearly depreciation:

Yearly depreciation = (Original cost − Trade-in Value) ×
Number of miles driven in one year ÷
Number of miles of life expectancy

Find the yearly depreciation of each of the four trucks listed in the table.

TRUCK	ORIGINAL COST	TRADE-IN VALUE	NUMBER OF MILES DRIVEN IN ONE YEAR	NUMBER OF MILES OF LIFE EXPECTANCY	YEARLY DEPRECIATION
1	$21 800	$3 000	46 500 miles	250 000 miles	
2	$16 075	$2 650	51 200 miles	200 000 miles	
3	$28 900	$3 800	57 910 miles	300 000 miles	
4	$36 610	$4 350	60 080 miles	350 000 miles	

3. Heat transfer by conduction is a basic process in refrigeration. In a refrigeration system a condenser transfers heat by conduction. Refrigerant gas enters a condenser at a high temperature. Heat is absorbed by water surrounding the tubing which contains the gas and the gas is cooled.

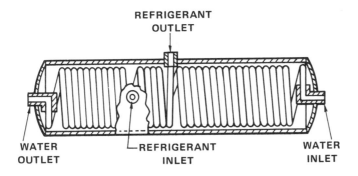

Find the rate at which heat is transferred by conduction in each of the following. Express the answers to the nearer whole Btu/min.

$$H = \frac{K \times A \times TD}{60 \times d}$$

	CONDUCTIVITY OF METAL (K)	SURFACE AREA OF METAL (A)	TEMPERATURE DIFFERENCE (TD)	THICKNESS OF METAL (d)	NUMBER OF BTU TRANSFERRED PER MINUTE (H)
a.	2 900	7.5 sq ft	9	0.031 in	
b.	1 740	6.3 sq ft	12	0.062 in	
c.	408	14.6 sq ft	8	0.250 in	
d.	2 900	3.8 sq ft	6.5	0.125 in	
e.	1 740	5.4 sq ft	10.5	0.093 in	

4. This problem deals with heat transfer by conduction. A technician wishes to determine the surface area of metal (A) required to transfer 120 000 Btu per minute (H) where $K = 1\,740$, $TD = 95°F$, and $d = 0.093$ inch.

$$A = \frac{60 \times H \times d}{K \times TD}$$

Find A in square feet to two decimal places.

5. Children's medicine doses are based on age or weight. A child's dose is only a portion of an adult dose. The following formula called Young's Rule is based on children's ages.

Child's dose = $\dfrac{\text{Age of child}}{\text{Age of child} + 12}$ × Average adult dose

Determine the child's dose, to one decimal place, for the ages listed in the table.

	AGE OF CHILD	AVERAGE ADULT DOSE	CHILD'S DOSE
a.	2	50 milligrams	
b.	3	30 milligrams	
c.	4	40 milligrams	
d.	3	60 milligrams	

6. A series-parallel electrical circuit is shown.

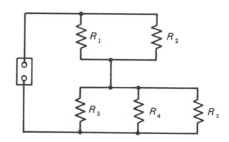

$$R_T = \dfrac{1}{\dfrac{1}{R_1} + \dfrac{1}{R_2}} + \dfrac{1}{\dfrac{1}{R_3} + \dfrac{1}{R_4} + \dfrac{1}{R_5}}$$

Find R_T to one decimal place when:
a. R_1 = 5.20 ohms, R_2 = 2.3 ohms, R_3 = 0.8 ohm, R_4 = 3.4 ohms, R_5 = 0.7 ohm.
b. R_1 = 0.8 ohm, R_2 = 3.4 ohms, R_3 = 1.5 ohms, R_4 = 5.3 ohms, R_5 = 0.9 ohm.

7. A flat is to be ground on a 0.750-centimetre diameter hardened pin. Determine the depth of material to be removed to produce a flat which is 0.325 centimetre long. Express the answer to the nearer thousandth of a centimetre.

$$C = \dfrac{D}{2} - 0.5 \times \sqrt{4 \times \left(\dfrac{D}{2}\right)^2 - F^2}$$

C is the depth of material to be removed (depth of cut).
D is the diameter of the pin.
F is the length of the required flat.

8. Banks use the compound-interest method of computing interest on money deposited. Compound-interest is computed on the amount of money originally deposited and on accumulated interest earned. One method of computing compound interest annually is:

$$A = P \times \left(1.00 + \frac{R}{100}\right)^n$$

A is the amount of money in the bank at the end of a certain number (n) of years. P is the amount of money originally deposited in the bank. It is called the original principal. R is the rate of interest earned expressed as a percent. n is the number of years the original principal (P) is left in the bank.

The table lists different values of P, R, and n. Find each value of A. No withdrawals or deposits have been made during n number of years.

	ORIGINAL PRINCIPAL (P)	PERCENT RATE OF INTEREST (R)	NUMBER OF YEARS (n)	AMOUNT OF MONEY IN BANK (A)
a.	$1 050	6.2 percent	2 years	
b.	$2 700	6.5 percent	3 years	
c.	$4 400	7 percent	2 years	
d.	$6 250	7.3 percent	3 years	

9. The inside dimensions of a gas tube boiler are given in metres. A pipefitter must determine the approximate number of cubic metres (volume) of steam space in the boiler. Steam space is the space above the boiler water line.

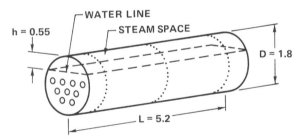

$$V = \frac{4 \times h^2}{3} \times \sqrt{\frac{D}{h} - 0.608} \times L$$

where V = number of cubic metres of steam space (volume)
h = height of steam space in metres
D = inside diameter of boiler in metres
L = inside length of boiler in metres

Find the number of cubic metres of steam space in the boiler to the nearer hundredth of a cubic metre.

10. Impedance is the total opposition to the flow of alternating current. It is the combined opposition of capacitive reactance (X_C), inductive reactance (X_L), and resistance (R). Impedance is measured in ohms. A circuit is shown.

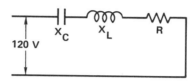

An electronics technician uses this formula to find the impedance (Z) in a series circuit.

$$Z = \sqrt{(X_L - X_C)^2 + R^2}$$

Determine the values of Z of circuits a, b, and c, in the table. Express the answers to the nearer whole ohm.

	INDUCTIVE REACTANCE (X_L)	CAPACITIVE REACTANCE (X_C)	RESISTANCE (R)	IMPEDANCE (Z)
a.	200 ohms	150 ohms	30 ohms	
b.	250 ohms	170 ohms	45 ohms	
c.	185 ohms	130 ohms	20 ohms	

SECTION 4
Percentage, Statistical Measures, and Graphs

UNIT 20
Percents and Simple Percentage Problems

OBJECTIVES

After studying this unit you should be able to
- Express decimal fractions and common fractions as percents.
- Express percents as decimal fractions and common fractions.
- Determine the percentage, given the base and rate.
- Determine the percent (rate), given the percentage and base.
- Determine the base, given the rate and percentage.

Each day, people are faced with various kinds of percentage problems to solve. Savings interest, loan payments, insurance premiums, and tax payments are based on percentage concepts.

Percentages are widely used in both business and nonbusiness fields. Merchandise selling prices and discounts, sales commissions, wage deductions, and equipment depreciation are determined by percentages. Business profit and loss are often expressed as percents. Percents are commonly used in making comparisons, such as production and sales increases or decreases over given periods of time. The basic percentage concepts have applications in many areas including business and finance, manufacturing, agriculture, construction, health, and transportation.

DEFINITION OF PERCENT

The *percent (%)* indicates the number of hundredths of a whole. This square is divided into 100 equal parts. The whole (large square) contains 100 small parts, or 100 percent of the small squares. Each small square is one part of 100 parts or 1/100 of the large square. Therefore, each small square is 1/100 of 100 percent or 1 percent.

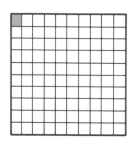

1 part of 100 parts

$\frac{1}{100} = 0.01 = 1\%$

Example: What percent of this square is shaded?

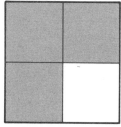

The large square is divided into 4 equal smaller squares. Three of the smaller squares are shaded.

3 parts of 4 parts

$\frac{3}{4} = 0.75 = 75\%$ *Ans*

Practice

A large square is divided into 25 equal smaller squares. Seven of the smaller squares are shaded. The 7 squares equal what percent of the whole large square?

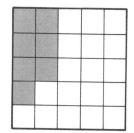

28% *Ans*

Exercise 20-1

Determine the percent of each figure that is shaded.

1.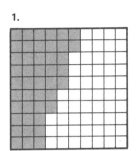
2.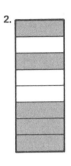
3.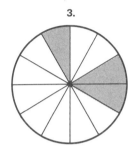
4.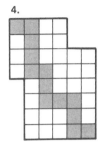

EXPRESSING DECIMAL FRACTIONS AS PERCENTS

A decimal fraction can be expressed as a percent by moving the decimal point 2 places to the right, and placing the percent symbol. (Moving the decimal point two places to the right is actually multiplying by 100.)

> Examples:
>
> 1. Express 0.015 2 as a percent.
>   Move the decimal point two places to the right.   0.01̬5 2 = 1.52% *Ans*
>   Place the percent symbol.
>
> 2. Express 3.876 as a percent.
>   Move the decimal point two places to the right.   3.87̬6 = 387.6% *Ans*
>   Place the percent symbol.

Practice

Express each decimal as a percent.

1. 0.135 . . . . 13.5%
2. 0.2 . . . . 20%
3. 0.53 . . . . 53%
4. 2.597 63 . . . . 259.763%
5. 3.7 . . . . 370%

EXPRESSING COMMON FRACTIONS AND MIXED NUMBERS AS PERCENTS

To express a common fraction as a percent, first express the common fraction as a decimal fraction. Then express the decimal fraction as a percent. If necessary to round, the decimal fraction must be two more decimal places than the desired number of places for the percent.

Examples:

1. Express $\frac{7}{8}$ as a percent.

    Express $\frac{7}{8}$ as a decimal fraction. $\frac{7}{8} = 0.875 = 87.5\%$ *Ans*

    Express 0.875 as a percent.

2. Express $5\frac{2}{3}$ as a percent to one decimal place.

    Express $5\frac{2}{3}$ as a decimal fraction. $5\frac{2}{3} = 5.667 = 566.7\%$ *Ans*

    Express 5.667 as a percent.

Practice

Express each common fraction as a percent. Round the answers to one decimal place when necessary.

1. $\frac{2}{5}$ 40%
2. $\frac{3}{4}$ 75%
3. $1\frac{3}{10}$ 130%
4. $2\frac{1}{6}$ 216.7%
5. $4\frac{1}{3}$ 433.3%

Exercise 20-2

Express each value as a percent.

1. 0.35   3. 0.04   5. 0.008   7. 2.076
2. 0.96   4. 0.062   6. 1.33   8. 0.063 9

9. 0.000 2   12. $\frac{3}{8}$ 15. $\frac{7}{16}$ 18. $2\frac{7}{25}$

10. 3.005   13. $\frac{3}{20}$ 16. $\frac{1}{250}$ 19. $4\frac{7}{8}$

11. $\frac{1}{4}$ 14. $\frac{4}{5}$ 17. $1\frac{1}{2}$ 20. $3\frac{1}{200}$

EXPRESSING PERCENTS AS DECIMAL FRACTIONS

Expressing a percent as a decimal fraction can be done by dropping the percent symbol and moving the decimal point two places to the left. Moving the decimal point two places to the left is actually dividing by 100.

Example: Express $37\frac{2}{5}\%$ as a decimal fraction.

Express $37\frac{2}{5}\%$ as 37.4% $37\frac{2}{5}\% = 37.4\% = 0.374$ *Ans*

Drop the percent symbol and move the decimal point two places to the left.

Practice

Express each percent as a decimal fraction. Round the answers to 3 decimal places.

1. 0.48% 0.005
2. $15\frac{3}{4}\%$ 0.158
3. 5% 0.050
4. 300% 3.000
5. $1\frac{1}{3}\%$ 0.013

EXPRESSING PERCENTS AS COMMON FRACTIONS

A percent is expressed as a fraction by first finding the equivalent decimal fraction. The decimal fraction is then expressed as a common fraction.

Example: Express 37.5% as a common fraction.

Express 37.5% as a decimal fraction. $37.5\% = 0.375 = \frac{375}{1\,000} = \frac{3}{8}$ *Ans*

Express 0.375 as a common fraction.

Practice

Express each percent as a common fraction.

1. 10% .... $\frac{1}{10}$

2. 3% .... $\frac{3}{100}$

3. $3\frac{1}{2}$% .... $\frac{7}{200}$

4. $222\frac{1}{2}$% .... $2\frac{9}{40}$

5. 0.5% .... $\frac{1}{200}$

Exercise 20-3

Express each percent as a decimal fraction or mixed decimal.

1. 82%
2. 19%
3. 3%
4. 2.6%
5. 27.76%
6. 103%
7. 224.9%
8. 0.6%
9. 4.73%
10. $12\frac{1}{2}$%
11. $\frac{3}{4}$%
12. 0.1%
13. $2\frac{3}{8}$%
14. 0.05%
15. $37\frac{1}{4}$%
16. $205\frac{1}{10}$%

Express each percent as a common fraction or mixed number.

17. 50%
18. 25%
19. 62.5%
20. 4%
21. 16%
22. 275%
23. 190%
24. 0.2%
25. 1.8%
26. 100.1%
27. 0.9%
28. 0.05%

TYPES OF SIMPLE PERCENTAGE PROBLEMS

A simple percentage problem has three parts. The parts are the rate, the base, and the percentage. In the problem, 10% of $80 = $8, the rate is 10%, the base is $80, and the percentage is $8. The *rate* is the percent. The *base* is the number of which the rate or percent is taken. It is the whole or a quantity equal to 100%. The *percentage* is the quantity of the percent of the base.

There are three types of simple percentage problems. The type used depends on which two quantities are given and which quantity must be found. The three types are as follows:
- Finding the percentage, given the rate (percent) and the base. A problem of this type is, "What is 15% of 384?"
- Finding the rate (percent), given the base and the percentage. A problem of this type is, "What percent of 48 is 12?"
- Finding the base, given the rate (percent) and the percentage. A problem of this type is, "50 is 30% of what number?"

FINDING THE PERCENTAGE, GIVEN THE BASE AND RATE

In some problems, the base and rate are given, and the percentage must be found. First, express the rate (percent) as an equivalent decimal fraction. Then multiply the base by the decimal fraction.

Examples:
1. What is 15% of 60?

   Express the rate, 15%, as an equivalent decimal fraction.   15% = 0.15

   Multiply the base, 60, by 0.15.   60 × 0.15 = 9 *Ans*

2. Find $6\frac{1}{4}$% of $180.

   Express the rate, 6 1/4%, as an equivalent $6\frac{1}{4}$% = 6.25% = 0.062 5
   decimal fraction.

   Multiply the base, $180, by 0.062 5 $180 × 0.062 5 = $11.25 *Ans*

Practice

1. An electrical contractor estimates the total cost of a wiring job as $3 200. Material cost is estimated as 35% of the total cost. What is the estimated material cost in dollars? Think the problem through to determine what is given and what is to be found. The base, $3 200, and the rate, 35%, are given. The percentage is to be found.

   $1 120 *Ans*

2. A retailer prices merchandise at 160% of the wholesale cost. What is the retail price of a piece of merchandise that has a wholesale cost of $422? Think the problem through to determine what is given and what is to be found. The base, $422, and the rate, 160%, are given. The percentage is to be found.

   $675.20 *Ans*

Exercise 20-4

Find each percentage. Round the answers to 2 decimal places when necessary.

1. 20% of 80
2. 2% of 80
3. 60% of 200
4. 15% of 150
5. 25% of 300
6. 7% of 140
7. 5% of 65
8. 0.8% of 214
9. 12.7% of 300
10. 22% of 13
11. 140% of 280
12. 1.8% of 1 000
13. 39% of 18.3
14. 0.42% of 50
15. 1% of 42.5
16. $8\frac{1}{2}$% of 375
17. $\frac{3}{4}$% of 132
18. 296.5% of 81
19. $15\frac{1}{4}$% of $35\frac{1}{4}$
20. $\frac{7}{10}$% of $139\frac{3}{10}$

21. A certain automobile cooling system has a capacity of 6 gallons. To give protection to −10°F, 40% of the cooling system capacity must be antifreeze. How many gallons of antifreeze should be used?

22. A print shop sells a used cylinder press for 42% of the original cost. If the original cost is $9 250, find the selling price of the used press.

23. A machine operator completes a job in 80% of the estimated time. The estimated time is 8 1/2 hours. How long does the job actually take?

24. The horsepower of an engine is increased by 8% after an engine is rebored. Find the horsepower increase if the engine is rated at 215 horsepower before it is rebored.

25. In an electrical circuit, a certain resistor takes 26% of the total voltage. The total voltage is 115 volts. Find how many volts are taken by the resistor.

26. A nurse computes a dosage of benadryl for a four year old child at 25% of the adult dosage. The adult dose is 50 milligrams. How many milligrams is the child's dose?

27. It is estimated that 37% of an apple harvest is spoiled by an early frost. Before the frost, the expected harvest was 3 800 bushels. How many bushels are estimated to be spoiled?

28. A floor area requires 320 board feet of lumber. In ordering material, an additional 12% is allowed for waste. How many board feet are allowed for waste?

FINDING THE PERCENT (RATE), GIVEN THE BASE AND PERCENTAGE

In some problems, the base and percentage are given, and the percent (rate) must be found. First, express the numbers as a fraction. The base is the denominator and the percentage is the numerator. Then, divide, and express the decimal fraction as its equivalent percent.

Examples:

1. What percent of 12 is 9.6?
   Express the numbers as a fraction. $\frac{9.6}{12}$
   Since a percent of 12 is to be taken, the base is 12.
   The base, 12, is the denominator. The percentage,
   9.6, is the numerator.
   Divide 9.6 by 12. $9.6 \div 12 = 0.8$
   Express 0.8 as a percent. $0.8 = 80\%$ Ans

2. What percent of 9.6 is 12?
   Express the numbers as a fraction. $\frac{12}{9.6}$
   Notice that although the numbers are the same as
   in Example 1, the base and percentage are reversed.
   Since a percent of 9.6 is to be taken, the base is 9.6.
   The base, 9.6, is the denominator. The percentage,
   12, is the numerator.
   Divide 12 by 9.6. $12 \div 9.6 = 1.25$
   Express 1.25 as a percent. $1.25 = 125\%$ Ans

3. An inspector rejects 23 out of a total production of 630 electrical switches. What percent of the total production is rejected?
   Think the problem through to determine what is
   given and what is to be found. Since a percent of
   the total production is to be taken, 630 is the base
   and 23 is the percentage. The percent (rate) is to
   be found.
   Express the numbers as a fraction. $\frac{23}{630}$
   The base, 630, is the denominator. The percentage,
   23, is the numerator.
   Divide 23 by 630. $23 \div 630 = 0.036\,5$
   (rounded to 4 decimal places)
   Express 0.036 5 as a percent. $0.036\,5 = 3.65\%$ Ans

Exercise 20-5

Find each percent (rate). Round the answers to 2 decimal places when necessary.

1. What percent of 8 is 4?
2. What percent of 20 is 5?
3. What percent of 100 is 37?
4. What percent of 84 is 70?
5. What percent of 70 is 84?
6. What percent of 258 is 97?
7. What percent of 132 is 206?
8. What percent of 19.5 is 5.5?
9. What percent of 1.25 is 0.5?
10. What percent of 0.5 is 1.25?
11. What percent of $6\frac{1}{2}$ is 2?
12. What percent of 134 is $156\frac{3}{4}$?
13. What percent of $\frac{7}{8}$ is $\frac{3}{8}$?
14. What percent of $\frac{3}{8}$ is $\frac{7}{8}$?
15. What percent of 3.08 is 4.76?
16. What percent of 0.65 is 0.09?

17. A garment requires 3 1/2 yards of material. If 1/4 yard of material is waste, what percent of the required amount is waste?

18. In making a 250 pound batch of bread dough, a baker uses 160 pounds of flour. What percent of the batch is made up of flour?

19. A casting, when first poured, is 17.875 centimetres long. The casting shrinks 0.188 centimetre as it cools. What is the percent shrinkage?

20. An electronics technician tests a resistor identified as 130 ohms. The resistance is actually 125 ohms. What percent of the identified resistance is the actual resistance?

21. A small manufacturing plant employs 130 persons. On certain days, 16 employees are absent. What percent of the total number of employees are absent?

22. The total amount of time required to machine a part is 12.5 hours. Milling machine operations take 7 hours. What percent of the total time is spent on the milling machine?

23. If 97 acres of 385 acres of timber are cut, what percent of the 385 acres is cut?

24. A road crew resurfaces 12.8 kilometres of a road which is 21.2 kilometres long. What percent of the road is resurfaced?

DETERMINING THE BASE, GIVEN THE PERCENT (RATE) AND THE PERCENTAGE

In some problems, the percent (rate) and the percentage are given, and the base must be found. First, express the percent as its decimal fraction equivalent. Then divide the percentage by the decimal fraction equivalent.

Examples:
1. 816 is 68% of what number?

   Express 68% as its equivalent decimal fraction. $68\% = 0.68$

   Divide the percentage, 816, by 0.68. $816 \div 0.68 = 1\,200$ *Ans*

   Note: The base, 1 200, is the whole or 100%.

2. $149.50 is 115% of what value?

   Express 115% as its equivalent decimal fraction. $115\% = 1.15$

   Divide the percentage, $149.50, by 1.15. $\$149.50 \div 1.15 = \$130$ *Ans*

   Note: The base, $130, is the whole or 100%.

Practice

A motor is said to be 80% efficient if the output (power delivered) is 80% of the input (power received). How many horsepower does a motor receive if it is 80% efficient and has an output of 6.2 horsepower (hp)? Think the problem through to determine what is given and what is to be found. The rate, 80%, and the percentage, 6.2 hp, are given. The base is to be found.

7.75 hp *Ans*

Note: The base, 7.75 hp, is the whole or 100%.

Exercise 20-6

Find each base. Round the answers to 2 decimal places when necessary.

1. 15 is 10% of what number?
2. 25 is 80% of what number?
3. 80 is 25% of what number?
4. 3.8 is 90% of what number?
5. 13.6 is 8% of what number?
6. 123 is 88% of what number?
7. 203 is 110% of what number?
8. $44\frac{1}{3}$ is 60% of what number?
9. $7\frac{1}{2}$ is 180% of what number?
10. 10 is $6\frac{1}{4}\%$ of what number?
11. 190.75 is 70% of what number?
12. 6.6 is 3.3% of what number?
13. 88 is 205% of what number?
14. 1.3 is 0.9% of what number?
15. $\frac{7}{8}$ is 175% of what number?
16. $\frac{1}{10}$ is $1\frac{1}{5}\%$ of what number?
17. On a production run, 8% of the units manufactured are rejected. If 120 units are rejected how many total units are produced?

18. An engine loses 4.2 horsepower through friction. The power loss is 6% of the total rated horsepower. What is the total horsepower rating?

19. This year's earnings of a company are 140% of last year's earnings. The company earned $910 000 this year. How much did the company earn last year?

20. An iron worker fabricates 70 feet of railing. This is 28% of a total order. How many feet of railing are ordered?

21. During a sale, 32% of a retailer's fabric stock is sold. The income received from the sale is $8 400. What is the total retail value of the complete stock?

22. A pump operating at 70% of its capacity discharges 4 200 litres of water per hour. When the pump is operating at full capacity, how many litres per hour are discharged?

23. How many pounds of mortar can be made with 75 pounds of hydrated lime if the mortar is to contain 15% hydrated lime?

24. The gasoline mileage of a certain automobile is 25% greater than last year's model. This represents an increase of 4.8 miles per gallon. Find the mileage per gallon of last year's model.

UNIT REVIEW

Exercise 20-7

Express each value as a percent.

1. 0.72
2. 0.05
3. 2.037
4. $\frac{1}{2}$
5. $\frac{1}{25}$
6. $1\frac{3}{8}$
7. 0.000 3
8. 3.190 6

Express each percent as a decimal fraction or mixed decimal.

9. 19%
10. 3.4%
11. 18.09%
12. 156%
13. 0.7%
14. $15\frac{1}{2}\%$
15. $\frac{3}{4}\%$
16. $310\frac{3}{10}\%$

Express each percent as a common fraction or mixed number.

17. 30%
18. 6%
19. 140%
20. 0.9%
21. 12.5%
22. 100.8%
23. 0.65%
24. 0.02%

Find each percentage. Round the answers to 2 decimal places when necessary.

25. 15% of 60
26. 3% of 40
27. 72% of 120
28. 5% of 240
29. 0.7% of 800
30. 42.6% of 500

31. 130% of 200
32. 210% of 8.5
33. $12\frac{1}{2}$% of 32
34. $\frac{1}{4}$% of 620
35. $10\frac{1}{10}$% of $92\frac{1}{5}$
36. $114\frac{3}{4}$% of 84

Find each percent (rate). Round the answers to 2 decimal places when necessary.
37. What percent of 10 is 2?
38. What percent of 2 is 10?
39. What percent of 88 is 22?
40. What percent of 275 is 108?
41. What percent of 53 is 77?
42. What percent of 2.84 is 0.8?
43. What percent of $12\frac{1}{4}$ is 3?
44. What percent of 312 is 400.9?
45. What percent of $\frac{3}{4}$ is $\frac{3}{8}$?
46. What percent of $13\frac{4}{5}$ is $6\frac{3}{10}$?

Find each base. Round the answers to 2 decimal places when necessary.
47. 20 is 60% of what number?
48. 60 is 20% of what number?
49. 4.1 is 25% of what number?
50. 340 is 150% of what number?
51. 44 is 73.5% of what number?
52. 9.3 is 238.6% of what number?
53. 0.84 is 2% of what number?
54. $20\frac{1}{2}$ is 71% of what number?
55. $\frac{3}{4}$ is 120% of what number?
56. $\frac{4}{5}$ is $3\frac{3}{5}$ of what number?

Find each percentage, percent (rate), or base. Round the answers to 2 decimal places when necessary.
57. What percent of 24 is 18?
58. What is 30% of 50?
59. What is 120% of 12.6?
60. 73 is 80% of what number?
61. What percent of $10\frac{1}{2}$ is 2?
62. ___?___ is 50% of 94.
63. 72% of 200 is ___?___.
64. What percent of 228 is 256?
65. 51 is 90% of what number?
66. 36.5 is ___?___ % of 27.
67. $2\frac{1}{4}$% of 150 is ___?___.
68. ___?___ is 18% of 120.
69. What percent of 36.2 is 45?
70. 16% of $9\frac{1}{4}$ is ___?___.
71. What is 33% of 93.6?
72. 550 is ___?___ % of 350.

PRACTICAL APPLICATIONS

Exercise 20-8
1. The carbon content of machine steel for gauges usually ranges from 0.15% to 0.25%.
    a. What is the minimum weight of carbon in 250 kilograms of machine steel?
    b. What is the maximum weight of carbon in 250 kilograms of machine steel?

2. A nautical mile is the unit of length used in sea and air navigation. A nautical mile is equal to 6 076 feet. What percent of a statute mile (5 280 feet) is a nautical mile?

3. Air is a mixture composed, by volume, of 78 percent nitrogen, 21 percent oxygen, and 1 percent argon.

   a. Find the number of cubic metres of nitrogen in 25 cubic metres of air.
   b. Find the number of cubic metres of oxygen in 25 cubic metres of air.
   c. Find the number of cubic metres of argon in 25 cubic metres of air.

4. An environmental systems technician estimates that weather-stripping the windows and doors of a certain building decreases the heat loss by 45% or 34 200 British thermal units per hour. Find the heat loss of the building before the weather stripping is installed.

5. A 30-ampere fuse carries a temporary 8% current overload. How many amperes of current flow through the fuse during the overload?

6. To increase efficiency and performance, the frontal area of an automobile is redesigned. The new design results in a decrease of frontal area to 73% of the original design. The frontal area of the new design in 18.25 square feet. What is the frontal area of the original design?

7. A 22-litre capacity radiator requires 6.5 litres of antifreeze to give protection to $-17°C$. What percent of the coolant is antifreeze?

8. A motor is 83% efficient; that is, the output is 83% of the input. What is the input if the motor delivers 10.25 horsepower?

9. When washed, a fabric shrinks 1/32 inch for each foot of length. What is the percent shrinkage?

10. How many pounds of butterfat are in 125 pounds of cream which is 34% butterfat?

11. An appliance dealer sells a television set at 170% of the wholesale cost. The selling price is $585. What is the wholesale cost?

12. A piece of machinery is purchased for $8 792. In one year the machine depreciates 14.5%. By how many dollars does the machine depreciate in one year?

13. An office remodeling job takes 5 3/4 days to complete. Before working on the job, the remodeler estimated the job would take 4 1/2 days. What percent of the estimated time is the actual time?

14. Engine pistons and cylinder heads are made of an aluminum casting alloy which contains 4% silicon, 1.5% magnesium, and 2% nickel.

    a. How many kilograms of silicon are needed to produce 500 kilograms of alloy?
    b. How many kilograms of magnesium are needed to produce 500 kilograms of alloy?
    c. How many kilograms of nickel are needed to produce 500 kilograms of alloy?

15. A transformer is rated at 320 watts. What percent of the rated power is lost by a 15-watt heat loss?

16. A batch of dough ferments for 3 hours, which is 80% of the total required fermentation time. Find the total required fermentation time.

17. A beautician sanitizes implements with a solution of disinfectant and water. A 4% solution contains 4% disinfectant and 96% water. How many ounces of solution are made with 2 ounces of disinfectant?

18. A beef stew recipe for 25 orders calls for 10 1/2 pounds of cubed beef. A chef wishes to increase the 25-order quantity to a 40-order quantity. The change requires a 60% increase of ingredients. How many additional pounds of cubed beef are required for the 40-order recipe?

19. A structural member is said to be in compression if a force or load tends to crush it. Compression values must be computed for various structural materials. Number 1 grade douglas fir has a compressive strength of 1 200 pounds per square inch. Western white pine has a compressive strength of 750 pounds per square inch. What percent of the compressive strength of western white pine is the compressive strength of douglas fir?

20. A fruit dealer considers probable spoilage in pricing merchandise. Figuring 12% spoilage, how many baskets of strawberries are spoiled out of a shipment of 450 baskets?

UNIT 21
More Complex Percentage Problems

OBJECTIVES

After studying this unit you should be able to

- Solve more complex percentage problems in which two of the three parts are not directly given.

In certain percentage problems, two of the three parts are not directly given. One or more additional operations may be required in setting up and solving a problem. Samples of these types of problems follow.

Example 1:

By replacing high-speed steel cutters with carbide cutters, a machinist increases production by 35%. Using carbide cutters, 270 pieces per day are produced. How many pieces per day were produced with high-speed steel cutters?

Think the problem through. The base (100%) is the daily production using high-speed steel cutters. Since the base is increased by 35%, the carbide cutter production of 270 pieces is 100% + 35% or 135% of the base. Therefore, the rate is 135% and the percentage is 270. The base is to be found.

Express 135% as its equivalent decimal fraction.   135% = 1.35

Divide the percentage, 270 pieces, by 1.35.   270 pieces ÷ 1.35 = 200 pieces *Ans*

Example 2:

A mechanic purchases a set of socket wrenches for $54.94. The purchase price is 33% less than the list price. What is the list price?

Think the problem through. The base (100%) is the list price. Since the base is decreased by 33%, the purchase price, $54.94, is 100% − 33% or 67% of the base. Therefore, the rate is 67% and the percentage is $54.94. The base is to be found.

Express 67% as its equivalent decimal fraction.   67% = 0.67

Divide the percentage, $54.94, by 0.67. $54.94 ÷ 0.67 = $82 *Ans*

Example 3:

An aluminum bar measures 137.168 millimetres before it is heated. When heated, the bar measures 137.195 millimetres. What is the percent increase in length? Express the answer to 2 decimal places.

Think the problem through. The base (100%) is the bar length before heating, 137.168 millimetres. The increase in length is 137.195 millimetres − 137.168 millimetres or 0.027 millimetre. Therefore, the percentage is 0.027 millimetre and the base is 137.168 millimetres. The rate (percent) is to be found.

Express the numbers as a fraction. The base, 137.168 mm, is the denominator. The percentage, 0.027 mm, is the numerator. $\dfrac{0.027 \text{ mm}}{137.168 \text{ mm}}$

Divide 0.027 mm by 137.168 mm.   0.027 mm ÷ 137.168 mm = 0.000 196 8

Express 0.000 196 8 as a percent.   0.000 196 8 = 0.019 68%
                                              (Rounded) 0.02% *Ans*

Example 4:

A manufacturing company produces 25 000 units in June. In July, 12.5% more units are produced than in June. The August production is 1 700 units less than the number of units manufacturered in July. What is the percent increase in August's production over June's production?

Think the problem through. Divide the problem into three basic steps: June − July production, July − August production, and June − August production. Find the number of units produced in July: June's production, 25 000 units, is the base (100%). The percent (rate) is 100% + 12.5% or 112.5%. Determine the percentage.

Express 112.5% as its decimal fraction equivalent.   112.5% = 1.125

Multiply the base, 25 000 units,   25 000 units × 1.125 = 28 125 units;
by 1.125.   28 125 units were produced in July.

Find the number of units produced in August. August production = July production − 1 700 units = 28 125 units − 1 700 units = 26 425 units. Compute the percent increase of August's production over June's production. Unit increase = 26 425 units − 25 000 units = 1 425 units. June's production, 25 000 units, is the base (100%). The unit production increase of August over June, 1 425 units, is the percentage. Determine the percent.

Express the numbers as a fraction.
The base, 25 000 units, is the denominator. $\frac{1\ 425}{25\ 000}$
The percentage, 1 425 units, is the numerator.

Divide 1 425 by 25 000.   1 425 ÷ 25 000 = 0.057
Express 0.057 as a percent.   0.057 = 5.7% *Ans*

Practice

A contractor is paid $95 000 for the construction of a building. The contractor's expenses are $40 200 for labor, $34 800 for materials, and $6 700 for miscellaneous expenses. What percent profit is made on this job?

Determine the amount of profit (percentage) by subtracting the total expenses from $95 000. Profit (percentage) = $95 000 − ($40 200 + $34 800 + $6 700) = $95 000 − $81 700 = $13 300. The base (100%) is $95 000. The percent profit is to be found.

14% Ans

PRACTICAL APPLICATIONS

Exercise 21-1

Solve each problem. Round the answers to 2 decimal places when necessary.

1. A vegetable farmer plants 87.4 acres of land this year. This is 15% more than the number of acres planted last year. Find the number of acres planted last year.

2. A laboratory technician usually prepares 250 millilitres of a certain solution. The technician prepares a new solution which is 35% less than that usually prepared. How many millilitres of new solution are prepared?

3. A mason is contracted to lay 125 feet of sidewalk. After laying 45 feet, what percent of the total job remains to be completed?

4. A printer has 800 reams of paper in stock at the beginning of the month. At the end of the first week, 28% of the stock is used. At the end of the second week, 50% of the stock remaining is used. How many reams of paper remain in stock at the end of the second week?

5. In the electrical circuit shown, the total current (total amperes) is equal to the sum of the individual currents. The total current is 10.15 amperes. What percent of the total current is taken by the refrigerator?

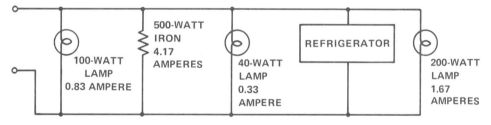

6. A welder estimates that 125 metres of channel iron are required for a job. Channel iron is ordered including an additional 20% allowance for scrap and waste. Actually, 175 metres of channel iron are used for the job. The amount actually used is what percent more than the estimated amount?

7. An alloy of red brass is composed of 85% copper, 5% tin, 6% lead, and zinc. Find the number of pounds of zinc required to make 450 pounds of alloy.

8. The day shift of a manufacturing firm produces 6% defective pieces out of a total production of 1 600 pieces. The night shift produces 4 1/2% defective pieces out of a total of 1 400 pieces. How many more acceptable pieces are produced by the day shift than the night shift?

226 Section 4 Percentage, Statistical Measures, and Graphs

9. Two pumps are used to drain a construction site. One pump, with a capacity of pumping 1 500 gallons per hour, is operating at 80% of its capacity. The second pump, with a capacity of pumping 1 800 gallons per hour, is operating at 75% of its capacity. Find the total gallons drained from the site when both pumps operate for 4 hours.

10. A resistor is rated at 2 500 ohms with a tolerance of ±6%. Tolerance is the amount of variation permitted for a given quantity. The resistor is checked and found to have an actual resistance of 2 320 ohms. By how many ohms is the resistor below the acceptable resistance low limit? Low limit = 2 500 ohms − 6% of 2 500 ohms.

11. A nurse is to prepare a 5% solution of sodium bicarbonate and water. A 5% solution means that 5% of the total solution is sodium bicarbonate. If 1 140 millilitres of water are used in the solution, how many grams of sodium bicarbonate are used?

    Note: Use 1 gram equal to 1 millilitre.

12. Forty-two grams of a certain breakfast cereal provides 20% of the United States recommended daily allowance of Vitamin D. Find the number of grams of cereal required to provide 90% of the recommended daily allowance of Vitamin D.

13. A contractor receives $122 000 for the construction of a building. Total expenses amounted to $110 400. What percent of the $122 000 received is profit?

14. A chef for a food catering service prepares 66 pounds of beef. The amount prepared is 20% more than is consumed. Find the number of pounds of beef consumed.

15. Before starting two jobs, a plumber has an inventory of eighteen 10-foot lengths of copper tubing. The first job requires 30% of the inventory. The second job requires 25% of the inventory remaining after the first job. How many feet of tubing remains in inventory at the end of the second job?

16. A baker usually prepares a 120-pound daily batch of dough. The baker wishes to reduce the batch by 20%. Find the number of pounds of dough prepared.

17. An alloy of stainless steel contains 73.6% iron, 18% chromium, 8% nickel, 0.1% carbon and sulphur. How many pounds of sulphur are required to make 5 000 pounds of stainless steel?

18. A homeowner computes fuel oil consumption during the coldest 5 months of the year as follows: November, 125 gallons; December, 145 gallons; January, 180 gallons; February, 165 gallons; March, 140 gallons. What percent of the 5-month consumption of fuel oil is January's consumption?

19. Two machines together produce a total of 2 015 pieces. Machine A operates for 6 1/2 hours and produces an average of 170 pieces per hour. Machine B operates for 7 hours. What percent of the average hourly production of Machine A is the average hourly production of Machine B?

20. A resistor is rated at 3 200 ohms with a tolerance of ±4%. The resistor is checked and found to have an actual resistance of 3 020 ohms. By what percent is the actual resistance below the low tolerance limit?

UNIT 22
Statistical Measures

OBJECTIVES

After studying this unit you should be able to

- Find the mean, median, and mode (measures of central tendency) in given exercises.
- Find the mean, median, and mode in practical applied problems.
- Find the range and mean deviation (measures of dispersion) in given exercises.
- Find the range and mean deviation in practical applied problems.

Statistics is the collection, organization, and use of numerical data. Government, industry, and business use statistics and statistical methods. Electronic computers, which are able to process great amounts of data quickly, are widely used in the production of statistics. Statistical methods range from simple to very advanced mathematical concepts.

In this text, only simple methods and their applications are presented. Two basic types of statistical measures are measures of *central tendency* of data and measures of *dispersion* or *spread* of data. The three commonly used measures of central tendency are the mean, the median, and the mode. Two useful measures of dispersion are the range and the mean deviation.

MEAN (AVERAGE)

The *mean* is the arithmetic average of a set of numbers. The mean measurement is the sum of the measurements divided by the number of measurements. Mean measurements are used for various kinds of measurements, such as length, time, volume, weight, and pressure. Mean values help to show the quality of products and manufacturing methods and are used to reduce measurement error.

Example: A steel fabrication firm keeps records of times taken to complete jobs. The records show the following times are taken to make 5 identical storage tanks: 57 hours, 61 hours, 55 hours, 49 hours, and 53 hours. To determine the labor and machine cost of a new order of tanks, an estimator computes the mean time of the previous jobs. Find the mean time.

Add the times of the five previous jobs.

```
  57 h
  61 h
  55 h
  49 h
+ 53 h
 275 h
```

Divide the sum, 275 h, by 5.   275 h ÷ 5 = 55 h *Ans*

Practice

To reduce measurement error, a tool and die maker makes four micrometer measurements of a die insert. The measurements are 23.178 millimetres, 23.180 millimetres, 23.182 millimetres, and 23.180 millimetres. What is the mean measurement?

23.180 mm *Ans*

Exercise 22-1

Find the mean for each set of numbers.

1. 15, 21, 17, 14, 25, 20
2. 8, 7, 4, 6, 8, 10, 9, 5, 6, 10, 9, 6
3. 61.5, 59.3, 58.0, 60.5, 64.3, 61.3, 62.1
4. 185.60, 178.35, 180.50, 179.32, 176.68
5. $12\frac{1}{2}, 11\frac{7}{8}, 12\frac{1}{16}, 11\frac{3}{4}, 11, 12\frac{7}{16}$
6. A meteorologist (weather forecaster) records hourly temperatures during a 12-hour period.

TIME	TEMPERATURE	TIME	TEMPERATURE	TIME	TEMPERATURE
6:00 AM	35° F	10:00 AM	46° F	2:00 PM	53° F
7:00 AM	38° F	11:00 AM	49° F	3:00 PM	53° F
8:00 AM	42° F	12:00 noon	49° F	4:00 PM	51° F
9:00 AM	44° F	1:00 PM	50° F	5:00 PM	50° F
				6:00 PM	48° F

a. Find, to the nearer degree, the mean temperature during the period from 6:00 AM to 12 noon.
b. Find, to the nearer degree, the mean temperature during the period from 12 noon to 6:00 PM.
c. Find, to the nearer degree, the mean temperature during the entire 12-hour period.

7. During one week, a truck-driver logs the following daily mileage: 373 miles, 458 miles, 290 miles, 387 miles, 412 miles. Find the average daily mileage during the week.

8. A sheet metal technician measures the thickness of six aluminum sheets as follows: 2.28 millimetres, 2.25 millimetres, 2.27 millimetres, 2.26 millimetres, 2.30 millimetres, 2.26 millimetres. Find the average sheet thickness.

9. Measure each of the ten lengths to the nearer sixteenth inch. Find the mean measurement of the lengths to the nearer sixteenth inch.

10. A chef trims excess fat waste from eight prime ribs of beef. The table lists the rib weights before and after trimming. Find the average weight of trim waste for the eight ribs.

	Weight Before Trimming	Weight After Trimming		Weight Before Trimming	Weight After Trimming
#1	24 oz	21 oz	#5	30 oz	24 oz
#2	27 oz	22 oz	#6	28 oz	24 oz
#3	20 oz	18 oz	#7	25 oz	23 oz
#4	18 oz	15 oz	#8	20 oz	17 oz

11. The inspection department of a manufacturing firm inspects 800 pieces of a product per week. In seven weeks production, the number of pieces rejected are 40, 25, 32, 81, 54, 16, and 27. Find the mean percent of rejection over the seven-week period. Express the answer to 2 decimal places.

MEDIAN

The *median* of a set of numbers is the middle number when the numbers are arranged in order of size. If the number of values is even, the median is the mean or average of the two middle numbers. The numbers may be arranged in either ascending order or descending order. In ascending order, the smallest number is written first and the largest number last. In descending order, the largest number is written first and the smallest number last.

Example: A technician checks 8 parts and records the following measurements: 5.17 cm, 5.20 cm, 5.15 cm, 5.19 cm, 5.16 cm, 5.21 cm, 5.16 cm, and 5.22 cm. Find the median.

Arrange the numbers in ascending order.

Since there is an even number of values, the median is the mean of the two middle values.

$$\frac{5.17 \text{ cm} + 5.19 \text{ cm}}{2} = 5.18 \text{ cm}$$

5.15 cm
5.16 cm
5.16 cm
⟶ 5.17 cm
⟶ 5.19 cm
5.20 cm
5.21 cm
5.22 cm

5.18 cm *Ans*

Practice

Determine the median value of the following weights. 12.5 lb, 11.6 lb, 13.6 lb, 12.2 lb, 12.0 lb, 13.4 lb, 11.8 lb, 11.6 lb, 12.3 lb

12.2 lb *Ans*

Exercise 22-2

Find the median for each set.

1. 18, 16, 21, 19, 18, 15, 17, 19, 22, 16, 20
2. 6.5, 5.3, 7.1, 8.7, 6.7, 5.2, 6.5, 7.0
3. 35.07, 34.98, 29.86, 28.93, 33.87, 32.35
4. 1 ft, $8\frac{1}{4}$ in, $10\frac{1}{2}$ in, $\frac{3}{4}$ ft, $\frac{2}{3}$ ft, $7\frac{3}{8}$ in, $\frac{7}{8}$ ft
5. 3.8 cm, 4.1 cm, 36 mm, 39 mm, 0.045 m, 4.2 cm

6. A construction engineering assistant made compression strength tests of 9 samples of cement used on a job. Find the median compression strength value.

Sample Number	Compression Strength (lb/sq in)	Sample Number	Compression Strength (lb/sq in)	Sample Number	Compression Strength (lb/sq in)
1	1 860	4	2 060	7	1 970
2	2 010	5	1 880	8	1 920
3	1 950	6	2 030	9	1 890

7. The number of units produced each month for one year by a garment manufacturer is shown in the table.

Month	Number of Units per Month	Month	Number of Units per Month	Month	Number of Units per Month
Jan	9 500	May	9 300	Sept	8 850
Feb	10 450	June	10 250	Oct	9 600
Mar	10 200	July	10 000	Nov	11 100
Apr	9 850	Aug	9 200	Dec	10 050

a. Find the median monthly production for January through July.
b. Find the median monthly production for the last eight months.
c. Find the median monthly production for the full year.

8. Find the median lot frontage of the six building lots shown. Express the answer to the nearer tenth metre.

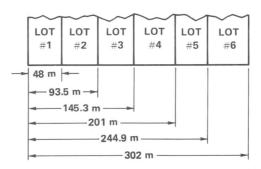

9. The blueprint dimension of the length of a part is given as 1.175 0" ± 0.000 8". Seven parts are made and measured. Find the median part length for the seven parts. Tolerance is the amount of variation permitted for a given quantity. High limit = 1.175 0" + 0.000 8". Low limit = 1.175 0" − 0.000 8".

Part #1: 0.000 3" under low limit
Part #2: on low limit
Part #3: 0.000 7" over high limit
Part #4: 0.000 2" over high limit
Part #5: 0.000 1" under low limit
Part #6: on high limit
Part #7: 0.000 5" under low limit

MODE

The *mode* of a set of numbers is the number that occurs most often. There may be more than one mode in a group of numbers.

Example: Determine the mode of the following values:

$2\frac{1}{2}$ min, 3 min, $4\frac{1}{2}$ min, 4 min, 2 min, 4 min, $2\frac{1}{2}$ min, $2\frac{3}{4}$ min, 4 min, and $4\frac{1}{2}$ min

Count the number of times that each value occurs.	4 min (3 times)
	$2\frac{1}{2}$ min (2 times)
Since 4 min occurs most often, 4 min is the mode.	All of the other values occur only 1 time.
	4 min *Ans*

It is helpful when dealing with many values, to make a frequency distribution table. A frequency distribution table lists the values and the number of times (frequency) that each value occurs.

Example: A technician tests the tensile strength of two castings from a production run of castings three times a day. This table lists the tensile strengths in pounds per square inch of tested castings for one week. What is the tensile strength mode of the castings tested during the week?

	MON	TUES	WED	THURS	FRI
TENSILE STRENGTH FOR TEST PERIOD 1 (lb/sq in)	2 400	2 375	2 350	2 425	2 500
	2 350	2 500	2 425	2 375	2 400
TENSILE STRENGTH FOR TEST PERIOD 2 (lb/sq in)	2 375	2 425	2 400	2 525	2 375
	2 425	2 375	2 425	2 325	2 325
TENSILE STRENGTH FOR TEST PERIOD 3 (lb/sq in)	2 500	2 525	2 325	2 350	2 325
	2 375	2 425	2 375	2 500	2 425

List the data in a frequency distribution table.

Both 2 425 lb/sq in and 2 375 lb/sq in occur the most number of times, 7. The two modes are 2 425 lb/sq in and 2 375 lb/sq in.

TENSILE STRENGTH (lb/sq in)	FREQUENCY (Number of times each value occurs)
2 525	2
2 500	4
2 425	7
2 400	3
2 375	7
2 350	3
2 325	4
Total	30

2 425 lb/sq in and 2 375 lb/sq in *Ans*

Exercise 22-3

Find the mode for each set of numbers.

1. 26, 17, 19, 20, 19, 16, 25, 19, 20, 22
2. 123.7, 119.6, 118.4, 117.6, 127.3, 132.5, 119.6, 120.9
3. $6\frac{1}{8}, 5\frac{2}{3}, 7\frac{3}{16}, 8\frac{1}{4}, 5\frac{5}{32}, 5\frac{2}{3}, 7\frac{1}{2}, 6\frac{3}{4}, 8\frac{5}{8}, 6\frac{1}{8}, 7\frac{7}{16}$
4. 0.023, 0.025, 0.019, 0.022, 0.018, 0.025, 0.021, 0.019, 0.024

5. An inspector of an electrical control manufacturing company tests thermostatic control switches. Each hour a sample of 10 switches is tested. Make a frequency distribution table and determine the mode for the given data.

TESTING TIME	TEMPERATURES AT WHICH THERMOSTATS SWITCHED ON (TEMPERATURES ARE IN DEGREES CELSIUS)				
8:00 AM	18.6	18.4	19.2	18.8	18.4
	19.8	19.0	18.4	18.0	19.0
9:00 AM	18.8	18.8	18.2	18.4	18.6
	19.4	18.4	18.8	19.2	18.8
10:00 AM	19.2	18.8	19.0	18.2	19.2
	18.4	18.6	19.6	19.2	19.6
11:00 AM	19.6	18.0	18.2	18.4	19.6
	19.4	18.8	18.0	19.0	18.6

6. A time study technician observes and times an assembly operation. Three different assemblers are timed. The observed times are recorded.

	ASSEMBLY TIMES IN MINUTES						
Assembler 1	3.4	3.0	3.5	3.1	3.4	3.0	3.1
	3.7	3.2	3.4	3.6	3.8	3.7	3.2
Assembler 2	3.0	2.9	2.7	3.1	2.6	2.7	2.9
	2.8	2.8	3.2	3.0	2.9	3.0	2.6
Assembler 3	3.1	3.2	2.9	3.2	3.3	2.9	3.4
	2.8	3.3	3.0	2.8	2.9	2.7	2.9

a. Find the mode time of each individual assembler.
b. Make a frequency distribution table of the complete data and find the mode of all of the times.

RANGE

The *range* is the difference between the highest and lowest numbers of a set of numbers.

> Example: Determine the range of the following resistances. 2 000 ohms, 2 023 ohms, 2 046 ohms, 1 970 ohms, 2 018 ohms, 1 972 ohms, 2 015 ohms, 1 980 ohms, 2 048 ohms, 2 020 ohms
>
> Subtract the lowest value, 1 970 ohms, from the highest value, 2 048 ohms.
>
> $$\begin{array}{r} 2\ 048\ \text{ohms} \\ -\ 1\ 970\ \text{ohms} \\ \hline 78\ \text{ohms}\ Ans \end{array}$$
>
> The range of values is 78 ohms.

MEAN DEVIATION

The *mean deviation* is the average or mean of the differences between the mean measurement and each of the individual measurements. To find the mean deviation, first find the mean. Then find the difference between the mean and each measurement. Last, find the sum of the differences and divide by the number of measurements.

> Example: A quality control technician measures five pieces as a sample of a production run. The pieces measure 3.125 3 cm, 3.124 7 cm, 3.125 2 cm, 3.124 8 cm, and 3.125 5 cm. Find the mean deviation. Round the answer to 4 decimal places.
>
> Find the mean.
>
> $$\begin{array}{r} 3.125\ 3\ \text{cm} \\ 3.124\ 7\ \text{cm} \\ 3.125\ 2\ \text{cm} \\ 3.124\ 8\ \text{cm} \\ +\ 3.125\ 5\ \text{cm} \\ \hline 15.625\ 5\ \text{cm} \end{array}$$
>
> 15.625 5 cm ÷ 5 = 3.125 1 cm
>
> Find the difference between the mean and each measurement.
>
> 3.125 3 cm − 3.125 1 cm = 0.000 2 cm
> 3.125 1 cm − 3.124 7 cm = 0.000 4 cm
> 3.125 2 cm − 3.125 1 cm = 0.000 1 cm
> 3.125 1 cm − 3.124 8 cm = 0.000 3 cm
> + 3 125 5 cm − 3.125 1 cm = 0.000 4 cm
> 　　　　　　　　　　　　　 0.001 4 cm
>
> Find the sum of the differences.
>
> Divide the sum, 0.001 4 cm by the number of measurements, 5.
>
> 0.001 4 cm ÷ 5 = 0.000 28 cm
>
> Round 0.000 28 cm to 4 decimal places. The mean deviation is 0.000 3 cm. This means that, on the average, the measurements vary 0.000 3 centimetre from the mean.
>
> 0.000 28 cm = 0.000 3 cm *Ans*

Exercise 22-4

Find both the range and the mean deviation for each set of numbers. Round the answers to 2 decimal places when necessary.

1. 12, 10, 8, 13, 10, 11, 8, 7, 12, 9
2. 6.12, 6.05, 6.18, 5.99, 5.97, 6.09, 6.11
3. 0.027, 0.033, 0.030, 0.028, 0.033, 0.029
4. $5\frac{5}{8}, 5\frac{3}{4}, 6\frac{1}{4}, 5\frac{7}{8}, 6, 6\frac{1}{8}$
5. A hydraulics technician checks six pressure line fittings to determine whether the fittings meet the minimum requirement of 1 250 pounds per square inch of pressure. The pressure readings are recorded as follows: 1 275 lb/sq in, 1 260 lb/sq in, 1 280 lb/sq in, 1 300 lb/sq in, 1 325 lb/sq in, and 1 270 lb/sq in. Find the range and mean deviation of the pressure readings.
6. At different times of day, the number of parts produced per minute by an automatic stamping machine are recorded. The following number of pieces per minute are recorded: 85, 87, 89, 92, 88, and 90. Find the range and mean deviation of the recorded parts.
7. A chemical laboratory assistant determines the amount of impurities in solutions. Five solutions, each containing 1 000 millilitres, are analyzed. Find the range and mean deviation of impurity, in millilitres, of the five solutions.
   Solution #1, 99.92% pure
   Solution #2, 99.86% pure
   Solution #3, 99.90% pure
   Solution #4, 99.88% pure
   Solution #5, 99.94% pure

UNIT REVIEW

Exercise 22-5

1. Find the mean of 16, 18, 22, 15, 17, 20, and 18.
2. Find the median of 2.7, 2.5, 3.0, 2.8, 2.6, 2.9, and 3.1.
3. Find the median of 0.75, 0.83, 0.79, 0.68, 0.85, and 0.77.
4. Find the mode of 13.5, 9.7, 11.0, 8.6, 10.2, 9.7, 8.6, 11.5, 9.7, 12.2, and 10.7.
5. Find the mean, median, and mode of 23.6, 22.8, 23.0, 24.5, 25.3, 24.3, 23.2, and 24.5.
6. Find the mean, median and mode of 0.023 5, 0.022 8, 0.024 1, 0.023 8, 0.023 5, 0.022 8, and 0.022 6.

PRACTICAL APPLICATIONS

Exercise 22-6

1. A small appliance dealer conducts a 5-day sale. The dollar amounts of appliances sold each day are $975, $1 142, $1 350, $1 508, and $1 262. Find the average daily dollar amount of sales over the 5-day period.

2. An automobile is tested for stopping distance at speeds of 60 miles per hour and 80 miles per hour. Six braking trials are made at each speed.

SPEED	STOPPING DISTANCE IN FEET					
60 mi/h	148	150	146	154	156	152
80 mi/h	265	272	268	273	276	266

   a. Find the mean stopping distance from 60 mi/h.
   b. Find the mean stopping distance from 80 mi/h.

3. Six samples of each of 3 operators hourly production of the same part are inspected. An inspector records the measurements.

MEASUREMENTS OF PIECES MADE BY EACH OPERATOR						
Operator 1	57.35 mm	57.32 mm	57.33 mm	57.29 mm	57.30 mm	57.28 mm
Operator 2	57.27 mm	57.32 mm	57.28 mm	57.31 mm	57.30 mm	57.26 mm
Operator 3	57.38 mm	57.35 mm	57.31 mm	57.29 mm	57.36 mm	57.33 mm

   Find the median measurement of the pieces made by the operators listed.
   a. Operator 1   c. Operator 3
   b. Operator 2   d. All three operators.

4. Inside caliper measurements, in inches, are made between sets of two consecutive holes as shown. The diameter of each hole is also given. Find the median center-to-center distance between two consecutive holes of the five center-to-center hole distances.

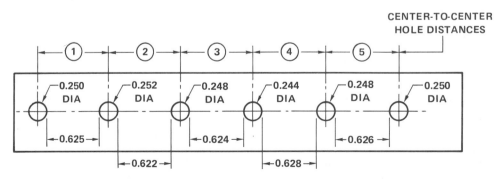

5. A company requires that persons applying for certain jobs take a mathematics ability test. Ability tests are given on three different days. Each day 16 applicants take the test. Test scores for all applicants during the 3 days are shown in the table.

TEST DAY	MATHEMATICS ABILITY TEST SCORES							
1	82	74	83	72	82	90	67	76
1	65	89	82	92	70	76	80	85
2	67	78	70	82	75	79	81	83
2	91	64	85	71	78	84	68	80
3	72	84	70	85	64	75	82	79
3	82	73	90	84	91	81	93	62

  a. Find the mode score for each day.
  b. Find the median score for each day.
  c. Make a frequency distribution table of the scores of all three days and find the mode and median of all scores.

6. Find both the range and the mean deviation for each set of numbers.
  a. 6, 5, 9, 4, 8, 11, 10, 7
  b. 0.009 8, 0.008 7, 0.009 2, 0.008 7, 0.009 8, 0.008 4

7. The hourly temperatures over a five-hour period are 65°F, 66°F, 70°F, 68°F, 66°F. Find the range and mean deviation of the recorded temperatures.

8. Six different metal alloy bars are tested to obtain expansion data. Each bar is 50.500 millimetres long before heating. All bars are heated to the same temperature. The lengths of the bars when heated are 50.502 mm, 50.501 mm, 50.504 mm, 50.503 mm, 50.505 mm, and 50.502 mm. Find the range and mean deviation of expansion of the bars. Express the mean deviation to 3 decimal places.

UNIT 23
Bar Graphs

OBJECTIVES

After studying this unit you should be able to
- Read and interpret data from given vertical and horizontal bar graphs.
- Draw and label vertical and horizontal bar graphs using given data.

USES OF GRAPHS

A *graph* shows the relationship between sets of quantities in picture form. Graphs are widely used in business, industry, government, and scientific and technical fields. Newspapers, magazines, books, and manuals often contain graphs. Since they are used in both occupations and everyday living, it is important to know how to interpret and construct basic types of graphs.

Statistical data are often time-consuming and difficult to interpret. Graphs present data in simple and concise picture form. Data, when graphed, often can be interpreted more quickly and is easier to understand.

TYPES AND STRUCTURE OF GRAPHS

There are many kinds of graphs which are designed for special purpose applications. An understanding of basic graphs, such as bar graphs and line graphs, provides a background for the reading and construction of other more specialized graphs.

Graph paper, which is also called coordinate and cross-section paper, is used for graphing data. Cross-section paper is available in various line spacings. Paper with 5 or 10 equal spaces in a given length is generally used.

Bar graphs and line graphs contain two scales. The *horizontal scale* is called the *x-axis*. The *vertical scale* is called the *y-axis*. The

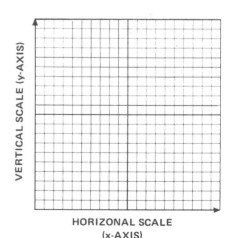

axes can be drawn at any convenient location, but are usually located on the bottom and left of the graph. The axes (scales) for all graphs in this unit are located at the bottom and left as shown.

A scale shows the values of the cross-sectional spaces. The scale values vary and depend on the data which are graphed.

READING BAR GRAPHS

On a bar graph, the lengths of the bars represent given data. To read a bar graph, first determine the value of each space on the scale (axis). If the bars are horizontal, determine the space value on the horizontal scale. If the bars are vertical, determine the space value on the vertical scale.

Next, locate the end of each bar. If the bar is horizontal, project down (vertically) to the horizontal scale. If the bar is vertical, project across (horizontally) to the vertical scale. Read each value on the appropriate scale. If the end of a bar is not directly on a line, estimate its value.

Example: This bar graph shows the monthly production of a manufacturing firm over a 5-month period.
     a. How many units are produced each month?
     b. What is the median monthly number of units produced?
     c. What is the mean monthly production?

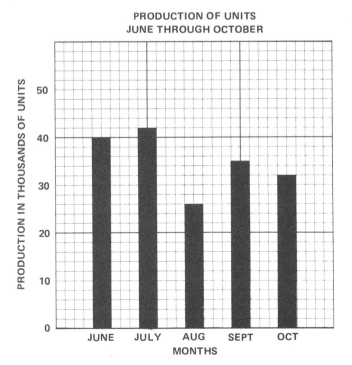

a. Find the vertical scale values. The major divisions represent 10 000 units. Each small space represents 10 000 units ÷ 5 = 2 000 units

From the end of each bar, project over to the vertical scale and read each value.

June: 40 000 *Ans*
July: 40 000 + (1 × 2 000) = 42 000 *Ans*
Aug: 20 000 + (3 × 2 000) = 26 000 *Ans*
Sept: 30 000 + (2.5 × 2 000) = 35 000 *Ans*
Oct: 30 000 + (1 × 2 000) = 32 000 *Ans*

b. Arrange the monthly values in ascending order.

26 000
32 000
⟶ 35 000
40 000
42 000

Find the middle value.

35 000 *Ans*

c. Find the sum of the units produced each month, and divide by the number of months.

175 000 ÷ 5 = 35 000 *Ans*

Bar graphs can be constructed so that each bar represents more than one quantity. Each bar is divided into two or more sections. To distinguish one section from another, sections are generally shaded, colored, or cross-hatched.

Example: This bar graph shows United States employment in major occupational groups for a certain year. Each group is represented by a bar. Each bar is divided into the number of males and females employed within the group.

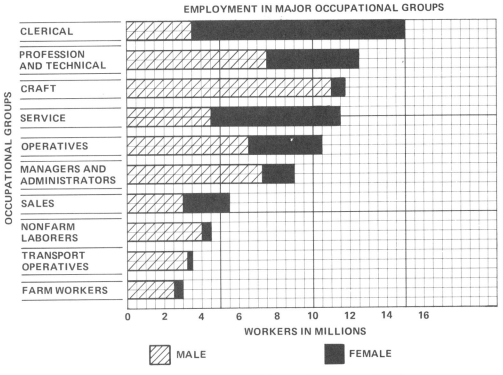

a. How many men are employed in service occupations?
b. How many women are employed in service occupations?
c. What percent, to the nearer whole percent, of the clerical group is made up of women?

Find the horizontal scale values. The numbered divisions each represent 2 000 000 workers. Each small space represents 2 000 000 ÷ 4 or 500 000 workers

a. Find the bar which represent service occupations

From the end of the male division of the bar, project down to the horizontal scale and read the value.

$$4\,000\,000 \text{ workers} + (1 \times 500\,000 \text{ workers}) =$$
$$4\,500\,000 \text{ workers or } 4.5 \text{ million workers}$$
in service occupations are men. *Ans*

b. From the end of the service occupations bar, project down to the horizontal scale and read the value.

$$10\ 000\ 000 \text{ workers} + (3 \times 500\ 000 \text{ workers}) =$$
$$11\ 500\ 000 \text{ workers or } 11.5 \text{ million}$$

Subtract the number of men from the total number of workers.

$$11\ 500\ 000 \text{ workers} - 4\ 500\ 000 \text{ workers} =$$
$$7\ 000\ 000 \text{ workers or 7 million}$$
in service occupations are women *Ans*

c. From the end of the clerical bar, project down to the horizontal scale. The total number of workers is found to be 15 million. The number of men equal 3 1/2 million. The number of women is found by subtracting the number of men from the total number of clerical workers. There are 11.5 million women in this group.

$$15\ 000\ 000 - 3\ 500\ 000 = 11\ 500\ 000 \text{ or } 11.5 \text{ million}$$

To find the percent of women, divide the number of women by the total number of workers in the group.

$$\frac{11.5 \text{ million}}{15 \text{ million}} = 0.766\ 6 = 77\% \text{ (to the nearer whole number)}$$
of the clerical group is made up of women *Ans*

Exercise 23-1

1. A firm's operating expenses for a certain year are shown on the horizontal bar graph.

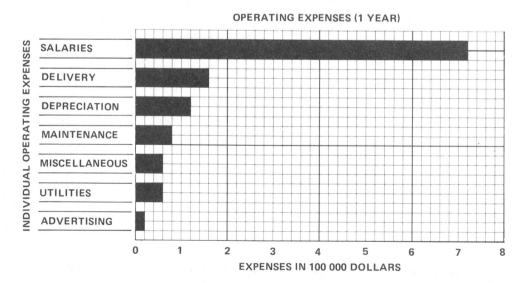

a. What is the amount of each of the seven operating expenses?
b. How many more dollars are spent for maintenance than for advertising?
c. What percent of the total operating expenses are paid for salaries? Express the answer to the nearer whole number.
d. The utilities expense shown is 20 percent greater than that of the previous year. How many dollars were spent on utilities during the previous year?

2. A vertical bar graph shows United States production of aluminum for each of eight consecutive years.

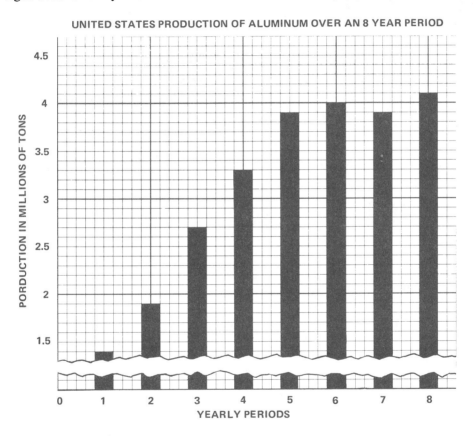

a. What is the number of tons of aluminum produced during each yearly period?
b. What is the mean (average) yearly tonnage produced during the eight year period?
c. How many more tons of aluminum were produced during the last year than during the first year of the eight-year period?
d. What is the percent increase in production of the last year over the first year of the eight-year period? Express the answer to the nearer whole percent.

3. The bar graph shows the motor vehicle production for a certain year by the six leading national producers.

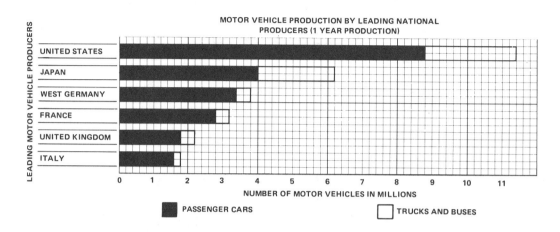

a. What is the total number of motor vehicles produced by the six nations?
b. What is the total number of trucks and buses produced by the six nations?
c. What percent of the total six nation motor vehicle production was produced by the United States? Express the answer to the nearer whole percent.

DRAWING BAR GRAPHS

The structure of bar graphs and how to read graph data have just been explained in this unit. Now it is possible to use what was learned to draw bar graphs from given data. Using graph paper saves time and increases the accuracy of measurements.

To draw bar graphs, first arrange the given data in a logical order. For example, group data from the smallest to the largest values or from the beginning to the end of a time period.

Next, decide which group of data is to be on the horizontal scale and which is to be on the vertical scale. Note: Generally, the horizontal scale is on the bottom and the vertical scale is on the left of the graph.

Now draw and label the horizontal and vertical scales. Note: Although the starting point is usually zero, any convenient value can be assigned.

Assign values to the spaces on the scales which conveniently represent the data. The data should be clear and easy to read.

Last, draw each bar to the required length according to the given data.

Example: A company's sales for each of eight years are listed. Indicate the year sales by a bar graph.

1977 – $2 400 000	1978 – $2 300 000	1974 – $1 800 000
1975 – $2 200 000	1976 – $2 100 000	1979 – $2 700 000
1980 – $2 900 000	1973 – $1 300 000	

Arrange the data in logical order. Rearrange the data in sequence from 1973 to 1980 as shown in this table.

YEAR	1973	1974	1975	1976	1977	1978	1979	1980
YEARLY SALES	$1 300 000	$1 800 000	$2 200 000	$2 100 000	$2 400 000	$2 300 000	$2 700 000	$2 900 000

Decide which data are to be on the horizontal scale and which on the vertical scale. It is decided to make the vertical scale the sales scale. The bars will be vertical. The vertical scale will be on the left and the horizontal scale on the bottom of the graph.

Draw and label the scales. The starting point will be zero, as shown.

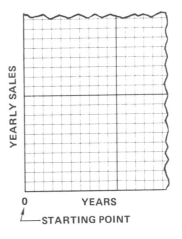

Assign values to scale spaces. Space years (centers of the bars) 5 small spaces apart. Assign each major division (10 small spaces) a value of $1 000 000. Label the major divisions 1, 2, and 3 to represent $1 000 000, $2 000 000, and $3 000 000 respectively. Each small space represents 0.1 million dollars or $100 000. For ease of reading, label each fifth small space in 0.5 million units.

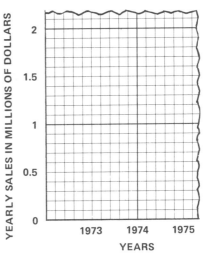

Draw each bar starting with 1973. The sales for 1973 were $1 300 000 or 1.3 million dollars. From the 1973 location on the horizontal scale, project up 1 major division (1 million) + 3 small spaces (0.3 million). Locate the endpoints of the remaining 7 bars the same way. Draw and shade each bar. The completed graph is shown below.

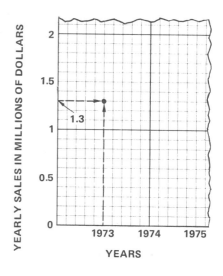

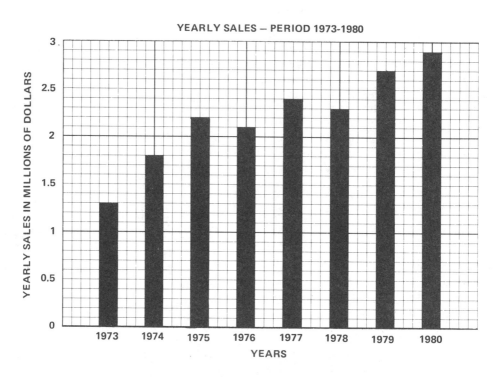

Exercise 23-2

Draw and label a bar graph for each of the following problems.

1. The table lists five sources of electrical energy in the United States. The percent of each source of the total energy production for a certain year is given. Draw a horizontal bar graph showing the table data.

ENERGY SOURCE	Coal	Gas	Oil	Hydropower	Nuclear
PERCENT OF TOTAL ENERGY	44%	21%	16%	15%	4%

2. Draw a vertical bar graph to show the dollar values of a chemical company's exports as listed in the table.

MONTH	Jan	Feb	Mar	April	May	June
MONTHLY EXPORTS	$850 000	$870 000	$1 100 000	$980 000	$1 090 000	$920 000

3. The table lists the six leading cattle producing states for a certain year. Draw a horizontal bar graph showing the table data.

STATE	Texas	Iowa	Kansas	Nebraska	Oklahoma	Missouri
NUMBER OF HEAD OF CATTLE	13 600 000	7 800 000	6 800 000	6 600 000	5 400 000	5 200 000

4. The United States nuclear power capacity in megawatts from 1975–1980 is shown. Draw a vertical bar graph showing the nuclear power capacity for the six years.

YEAR	1975	1976	1977	1978	1979	1980
NUCLEAR POWER CAPACITY IN MEGAWATTS	17 000	45 000	66 000	77 000	87 000	98 000

UNIT REVIEW

PRACTICAL APPLICATIONS

Exercise 23-3

1. The bar graph shows the quarterly production for a manufacturing firm over a period of a year.

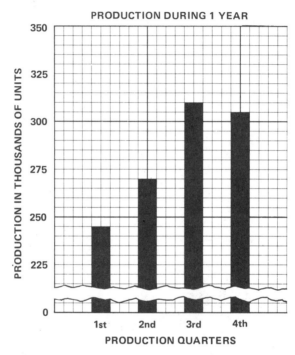

a. How many units are produced during each quarter?
b. What is the average (mean) quarterly production during the year?
c. What percent of the total yearly production is manufactured during the first half of the year? Express the answer to the nearer whole percent.

2. The number of domestic and international passengers during one year using six major international airports is shown.

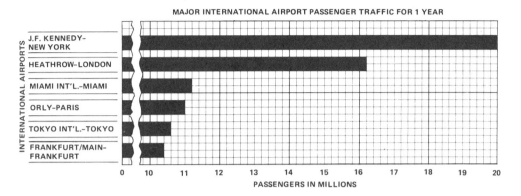

MAJOR INTERNATIONAL AIRPORT PASSENGER TRAFFIC FOR 1 YEAR

a. What is the total passenger traffic for the six airports?
b. What percent of the total passenger traffic of the six airports is the traffic through J. F. Kennedy International Airport? Express the answer to the nearer whole percent.
c. The number of passengers shown on the graph using Heathrow International Airport represents an eight percent increase above the previous year. How many passengers used Heathrow the previous year?

3. The production of five major United States farm products for a certain year is shown in the table. Draw and label a horizontal bar graph showing the table data.

CROP	Corn, Grain	Wheat	Soybeans	Oats	Barley
NUMBER OF BUSHELS PRODUCED IN BILLIONS OF BUSHELS	5.4	1.5	1.3	0.7	0.5

4. Draw a vertical bar graph showing the United States imports for eight consecutive years as shown in the following table.

YEAR	1st	2nd	3rd	4th	5th	6th	7th	8th
IMPORTS IN BILLIONS OF DOLLARS	18	21	25	26	33	36	39	45

5. A company's operating expenses for one quarter are as follows:

Salaries, $430 000	Maintenance, $80 000
Delivery, $150 000	Utilities, $70 000
Depreciation, $110 000	Advertising, $40 000

Draw and label a horizontal bar graph showing the quarterly operating expenses.

UNIT 24
Line Graphs

OBJECTIVES

After studying this unit you should be able to

- Read and interpret data from given broken-line, straight-line, and curved-line graphs.
- Draw and label broken-line, straight-line, and curved-line graphs by directly using given data.
- Draw and label straight-line and curved-line graphs by expressing given formulas as table data.
- Identify given table data in terms of constant or variable rates of change and identify the type of graph which would be produced by plotting the data.

Line graphs show changes and relationships between quantities. Line graphs are widely used to graph the following two general types of data: data where there is no causal relationship between quantities, and data where there is a causal relationship between quantities.

Data where there is *no* causal relationship between quantities. When the data are graphed, the graph shows a changing condition usually identified by a broken line. This type of graph is called a *broken-line graph*. The time and temperature graph shown is an example of a broken-line graph.

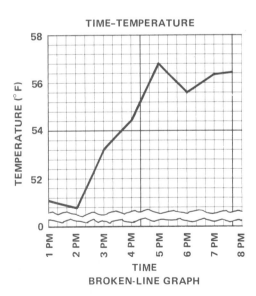

BROKEN-LINE GRAPH

Data where there *is* a causal relationship between quantities. The quantities are related to each other by a mathematical rule or formula. When the data are graphed, the line is usually a straight line or a smooth curve.

This graph is an example of a *straight-line graph*. The quantities are the perimeters of squares in relation to the lengths of their sides. The perimeter of a square is the distance around the square. The formula for the perimeter of a square is: Perimeter = 4 times the side length, $P = 4s$.

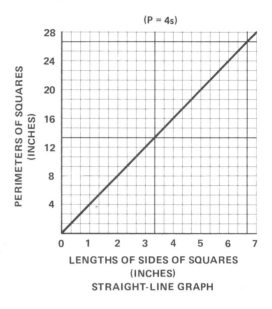

This graph is an example of a *curved-line graph*. The quantities are the areas of squares in relation to the lengths of their sides. The formula for the area of a square is: Area = the square of a side, $A = s^2$.

READING LINE GRAPHS

Information is read directly from a line graph by locating a value on one scale, projecting to a point on the graphed line, and projecting from the point to the other scale. More data can generally be obtained from a line graph than from a bar graph. Data between given scale values can be read. In most cases, the values read are close approximations of the true values.

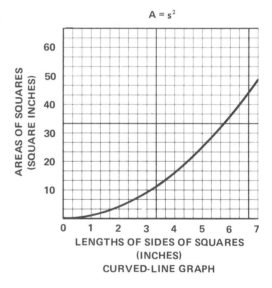

To read a line graph, first determine the value of each space on the scales. Then locate the given value on the appropriate scale. The value may be on either the horizontal or vertical scale depending on how the graph is organized. If the value does not lie directly on a line, estimate its location.

From the given value, project up from a horizontal scale or across from a vertical scale to a point on the graphed line. From the point, project across or down to the other scale. Read the scale value. If the value does not lie directly on the line, estimate the value.

Example: A quality control assistant constructs the broken-line graph shown. The percent defective pieces of the total production for each of 10 consecutive days is given.

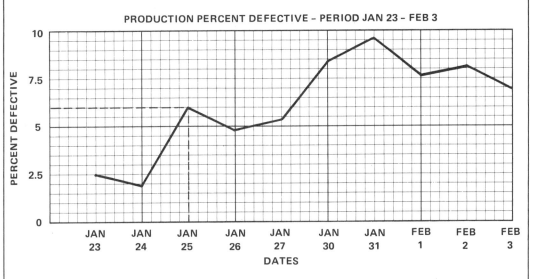

a. Find, to the nearer 0.5%, the percent defective pieces for January 25.
b. If the total production for January 25 is 1 550 pieces, how many pieces are defective?
c. Find the median daily percent defective during the 10-day period.

a. Find the vertical scale values. There is 2.5 percent between each numbered division. Each small space represents
$$2.5\% \div 5 = 0.5\%$$

Locate January 25 on the horizontal scale. Project up to a point on the graphed line. From this point, project across to the percent defective scale. Read the value to the nearer 0.5 percent.

(Jan 25) 6% *Ans*

b. Find 6% of 1 550 pieces.

$$0.06 \times 1\,550 = 93\ Ans$$

c. Find each daily percent.

Arrange in ascending order.

Find the average of the two middle values.

Jan 24 – 2.0%
Jan 23 – 2.5%
Jan 26 – 5.0%
Jan 27 – 5.5%
⟶ Jan 25 – 6.0%
⟶ Feb 3 – 7.0%
Feb 1 – 7.5%
Feb 2 – 8.0%
Jan 30 – 8.5%
Jan 31 – 9.5%

$$\frac{6.0\% + 7.0\%}{2} = 6.5\%\ Ans$$

256   Section 4 Percentage, Statistical Measures, and Graphs

READING COMBINED DATA LINE GRAPHS

Two or more sets of data are often combined on the same graph. Graphs of this type are useful in showing relationships and making comparisons between sets of data. Comparing information on two or more lines on a graph can be done more quickly and interpreted easier than comparing listed data.

Example: Acceleration tests are made for two cars. One car is a manufacturer's standard production model. The other car is a high-performance competition model. Acceleration data obtained from tests are plotted on this graph. Observe that gear shift points are also indicated on the graph.

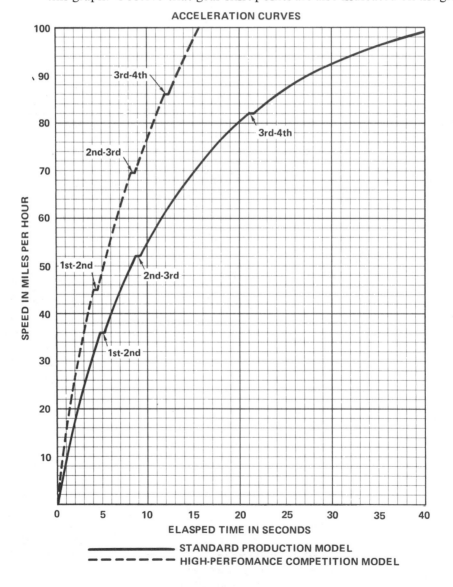

a. How many seconds are required for the high-performance model to accelerate from 0 to 60 miles per hour?
b. What is the speed of the production model after 22 seconds from a standing start?
c. How many seconds are required by the production model to accelerate from 40 to 70 miles per hour?
d. At the end of 12 seconds, how much greater is the speed of the high-performance model than the production model?
e. At what speeds were the shifts through gears made on the production model?

a. Locate 60 mi/h on the speed scale. Project across to a point on the graph for high-performance (broken line). Read the time to the nearer second.

7 seconds *Ans*

b. Locate 22 seconds on the time scale. Project up to a point on the graph for the production model (solid line). Read the speed.

82 mi/h *Ans*

c. Locate 40 mi/h on the speed scale. Project across to a point on the graph for the production model (solid line). The time is 6 seconds.

Locate 70 mi/h on the speed scale. Project across to a point on the graph for the production model (solid line). The time is 15 seconds.

Subtract. 15 sec − 6 sec = 9 sec *Ans*

d. Locate 12 seconds on the time scale. Project up and across for each model.

Subtract. 86 mi/h − 61 mi/h = 25 mi/h *Ans*

e. Locate on the graph for the production model each place of gear change. Read the speed for each gear change.

36 mi/h *Ans*
52 mi/h *Ans*
82 mi/h *Ans*

Section 4 Percentage, Statistical Measures, and Graphs

Exercise 24-1

1. Temperature in degrees Celsius for the different times of day are shown on the graph. Express the answers to the nearer 0.2 degree.

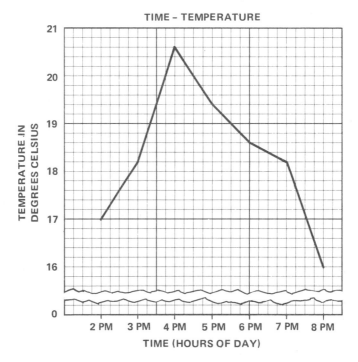

a. What is the temperature for each of the hours shown on the graph?
b. What is the mean hourly temperature during the six-hour period?
c. What is the temperature range during the six-hour period?

2. The surface or rim speed of a wheel is the number of feet that a point on the rim of the wheel travels in one minute. The surface speed depends on the size of the wheel diameter and the number of revolutions per minute (r/min) that the wheel is turning. The graph shows the surface speeds of different diameter wheels. All wheels are turning at 320 revolutions per minute. Express the answers for surface speeds to the nearer 10 feet per minute and diameters to the nearer 0.2 inch.

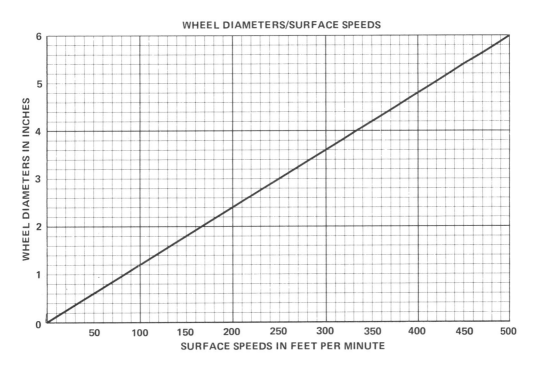

a. What is the surface speed of each of the wheel diameters shown on the graph?
b. What is the surface speed of a 2.6-inch diameter wheel?
c. What is the surface speed of a 4.2-inch diameter wheel?
d. What is the surface speed of a 5.8-inch diameter wheel?
e. What diameter wheels are needed to give 390 feet per minute surface speed?
f. What diameter wheels are needed to give 440 feet per minute surface speed?
g. What diameter wheels are needed to give 120 feet per minute surface speed?

3. An electrical current/resistance graph is shown. A constant power of 150 watts is being consumed. Express the answers for resistance to the nearer ohm and for current to the nearer 0.2 ampere.

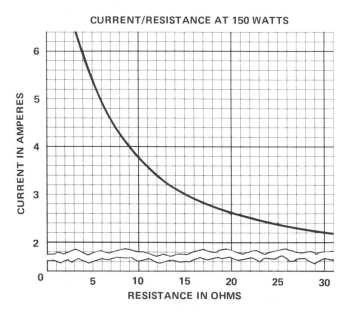

a. What is the current for each of the resistances shown on the graph?
b. What is the resistance for each of the currents shown on the graph? Disregard the 2 ampere current.
c. What is the resistance for a current of 2.4 amperes?
d. What is the resistance for a current of 4.8 amperes?
e. What is the resistance for a current of 5.6 amperes?

4. This graph shows the brake horsepower of two engines at various engine speeds. One engine is fitted with a medium compression head, the other with a high compression head. Express the answers to the nearer 5 brake horsepower or to the nearer 100 r/min.

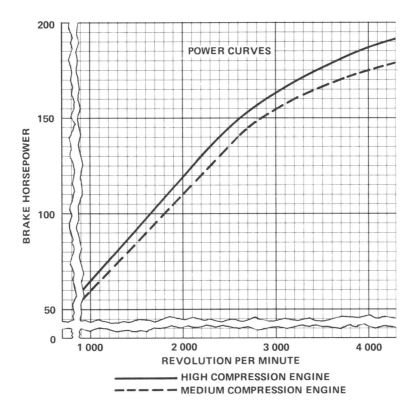

a. What is the brake horsepower for the medium compression engine at 2 200 r/min?
b. What is the brake horsepower for the high compression engine at 3 100 r/min?
c. What is the brake horsepower for the high compression engine at 3 800 r/min?
d. How many revolutions per minute are required for the medium compression engine when developing 140 brake horsepower?
e. How many revolutions per minute are required for the high compression engine when developing 160 brake horsepower?
f. How many revolutions per minute are required for the high compression engine when developing 185 brake horsepower?
g. What is the increase in brake horsepower of each engine when the engine speeds are increased from 2 200 r/min to 3 700 r/min?
h. How many brake horsepower greater is the high compression engine than the medium compression engine at 1 400 r/min?
i. How many brake horsepower greater is the high compression engine than the medium engine at 2 600 r/min?
j. How many brake horsepower greater is the high compression engine than the medium engine at 4 200 r/min?

262 Section 4 Percentage, Statistical Measures, and Graphs

DRAWING LINE GRAPHS

The first four steps of constructing a line graph are the same as for constructing a bar graph.

To draw a line graph, first arrange the given data in a logical order. For example, group data from the smallest to the largest values or from the beginning to the end of a time period. Note: Sometimes data is not given directly, but must be computed from other given facts, such as formulas.

Then decide which group of data is to be on the horizontal scale and which is to be on the vertical scale. Generally, the horizontal scale is on the bottom and the vertical scale is on the left of the graph.

Draw and label the horizontal and vertical scales. Next, assign values to the spaces on the scales which conveniently represent the data. The data should be clear and easy to read.

Plot each pair of numbers (coordinates). Project up from the horizontal scale and across from the vertical scale. Place a dot on the graph where the two projections meet.

Last, connect the plotted points with a straightedge or curve. Depending on the given data, the line may be straight, broken, or curved.

DRAWING BROKEN-LINE GRAPHS

Quantities that are not related to each other by a mathematical rule or formula form a broken-line graph.

Example: A company's profit for each of six weeks is listed in this table. Draw a line graph showing this data.

WEEK	1	2	3	4	5	6
WEEKLY PROFIT	$1 350	$1 100	$1 600	$1 850	$1 750	$1 900

Arrange the data in logical order. Since the data are listed in order from week 1 to week 6, no rearrangement is necessary.

Decide which data are to be on the horizontal scale and which on the vertical scale. It is decided to make weeks the horizontal scale. Make the weeks scale on the bottom and the profit scale on the left of the graph.

Draw and label the scales.

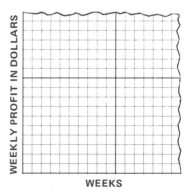

Assign values to scale spaces. Space each week 5 small spaces apart on the horizontal scale. On the vertical scale, start the numbering at $1 000. Observe there is no profit less than $1 000. Assign each major division (10 small spaces) a value of $500. Each small space represents $500 ÷ 10 or $50. Label each fifth space. Each fifth space represents 5 × $50 or $250.

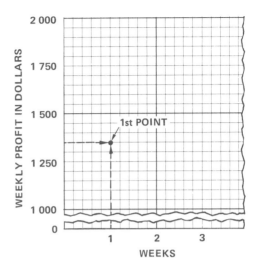

Plot each pair of values. From the week 1 location project up, and from $1 350 project across. Place a small dot where the projections meet. Locate the remaining 5 points the same way.

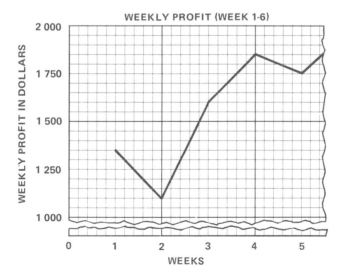

Connect the plotted points. Draw a straight line between each of two consecutive points.

Exercise 24-2

1. Draw a broken-line graph to show the six-month production of iron ore in the United States as listed in the table. Production is given in units of millions of metric tons.

MONTH	June	July	Aug	Sept	Oct	Nov
MONTHLY PRODUCTION OF IRON ORE IN MILLIONS OF METRIC TONS	6.2	6.4	5.9	5.7	6.0	6.1

2. The table lists the percent of defective pieces of the total number of pieces produced daily by a manufacturer. The data are recorded for seven consecutive working days. Draw and label a broken-line graph showing the table data.

DATE	March 27	March 28	March 29	March 30	March 31	April 3	April 4
DAILY PERCENT DEFECTIVE	3.2%	4.6%	6.0%	5.6%	6.8%	7.2%	4.4%

DRAWING STRAIGHT-LINE GRAPHS

Quantities that are related to each other by a mathematical rule or formula form a curved or straight line when graphed. If the two quantities change at a constant rate, a straight-line graph is formed.

Example: Draw a line graph for the perimeters of squares.

$$P = 4s$$

Use side lengths of 1 centimetre through 8 centimetres.

The data are not given directly.
The perimeters related to each of the sides must be computed.
Substitute each side length in the formula to determine the corresponding perimeter value.
Note the change in the perimeter is constant 4 cm for each centimetre change in the length of the side.

4×1 cm = 4 cm
4×2 cm = 8 cm
4×3 cm = 12 cm
4×4 cm = 16 cm
4×5 cm = 20 cm
4×6 cm = 24 cm
4×7 cm = 28 cm
4×8 cm = 32 cm

Organize the data in table form as shown.

LENGTHS OF SIDES IN CENTIMETRES	1	2	3	4	5	6	7	8
PERIMETERS OF SIDES IN CENTIMETRES (P = 4s)	4	8	12	16	20	24	28	32

This is the completed graph

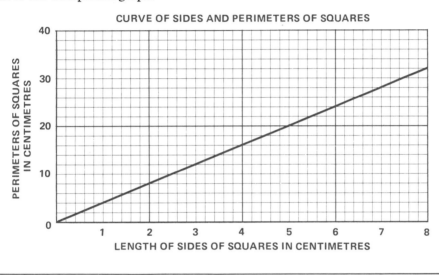

CURVE OF SIDES AND PERIMETERS OF SQUARES

DRAWING CURVED-LINE GRAPHS

The area of a square does not increase at a constant amount for each unit change in the length of the side. If the variation is not constant, a curved line is formed.

Example: Draw a line graph for the areas of squares.

$$A = s^2$$

Use side lengths of 1 centimetre through 8 centimetres.

The data are not given directly. The areas related to each of the sides must be computed. Substitute each side length in the formula to determine the corresponding area value. Note the change in the area is not a constant amount for each centimetre change in the length of the side.

1 cm × 1 cm = 1 cm²
2 cm × 2 cm = 4 cm²
3 cm × 3 cm = 9 cm²
4 cm × 4 cm = 16 cm²
5 cm × 5 cm = 25 cm²
6 cm × 6 cm = 36 cm²
7 cm × 7 cm = 49 cm²
8 cm × 8 cm = 64 cm²

Organize the data in table form as shown.

LENGTHS OF SIDES IN CENTIMETRES	1	2	3	4	5	6	7	8
AREAS IN SQUARE CENTIMETRES ($A = s^2$)	1	4	9	16	25	36	49	64

This is the completed graph.

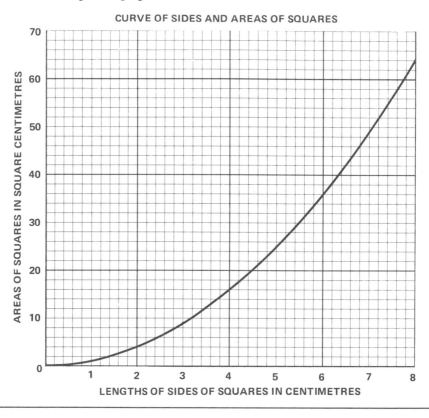

Exercise 24-3

1. The data in the table show electrical currents at various voltages. All listed currents are flowing through a constant resistance of 20 ohms. Draw and label a current-voltage graph using the table data.

VOLTAGE (Volts)	10	20	30	40	50
CURRENT (Amperes)	0.5	1.0	1.5	2.0	2.5

2. Draw and label an acceleration curve to show the acceleration data for a certain automobile as listed in the table. From a standing start, the car's speed is given for various elapsed time values.

ELAPSED TIME IN SECONDS	0	5	10	15	20	25	30
SPEED IN MILES PER HOUR	0	32	54	68	78	86	90

3. The rise and run of a roof are shown.

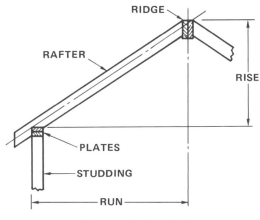

Rise = pitch × 2 × run

Draw a graph showing the relation of rise to run with a constant pitch value of 1/3. Use run values of 8', 12', 16', 20', 24', and 28'.

Note: It is helpful to make a table of the values to be graphed.

4. The area of a circle is approximately equal to 3.141 6 times the square of the radius.

$$A = 3.141\ 6 \times r^2$$

Draw a curved-line graph to show the relation of radii to areas. Use radii from 0.5 metre to 3.5 metres in 0.5 metre intervals. Label the radius scale in metres and the area scale in square metres.

Note: It is helpful to make a table of the values to be graphed.

5. The cutting speeds of different size (diameter) drills is shown in the table. All drills are turning at a constant 600 revolutions per minute. Copy the table and fill in the missing changes in cutting speed values. If the table data are plotted, is the graph a curved-line graph or a straight-line graph? Do not graph the table data.

DRILL DIAMETERS IN INCHES	0.250	0.500	0.750	1.000	1.250	1.500
CUTTING SPEEDS IN FEET PER MINUTE	39.25	78.50	117.75	157.00	196.25	235.50
CHANGES IN CUTTING SPEEDS IN FEET PER MINUTE						

6. Lengths of sides of cubes with their respective volumes are given in the table. Copy the table and fill in the missing changes in volume values. If the table data are plotted, is the graph a curved-line graph or a straight-line graph? Do not graph the table data.

LENGTHS OF CUBE SIDES IN METRES	0.4	0.8	1.2	1.6	2.0	2.4
VOLUMES OF CUBES IN CUBIC METRES	0.064	0.512	1.728	4.096	8.000	13.824
CHANGES IN VOLUMES IN CUBIC METRES						

UNIT REVIEW

PRACTICAL APPLICATIONS

Exercise 24-4

1. A wholesale distributor's monthly profits for six consecutive months are shown on the graph.

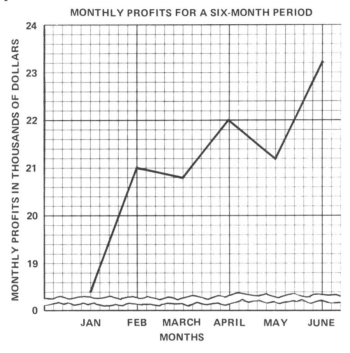

a. What is the profit for each of the six months? Express the answers to the nearer $200.
b. What is the average (mean) monthly profit during the six-month period?
c. What is the percent increase in profit in June over May? Express the answer to the nearer tenth percent.

2. The carburetor of an automobile engine mixes air with gasoline. There is an ideal mixture of gasoline and air which gives maximum fuel economy. The mixture of gasoline and air is called an air-fuel ratio. The number of pounds of air in the mixture is compared with the number of pounds of gasoline. The graph shows air-fuel ratios in relation to fuel consumption in miles per gallon of gasoline for a certain car. Express the answers to the nearer whole mile per gallon.

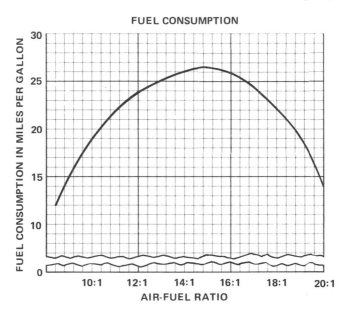

a. How many miles per gallon of gasoline are obtained at a 12:1 fuel-air ratio?
b. How many miles per gallon of gasoline are obtained at a 16:1 fuel-air ratio?
c. How many miles per gallon of gasoline are obtained at a 17.2:1 fuel-air ratio?
d. How many miles per gallon of gasoline are obtained at a 18.8:1 fuel-air ratio?
e. What fuel-air ratio gives the greatest number of miles per gallon?
f. How many more miles per gallon is obtained with a 14:1 air fuel ratio than with a 20:1 ratio?

3. The graph shows thickness of a certain alloy of aluminum sheet in relation to the weights of one square foot of material.

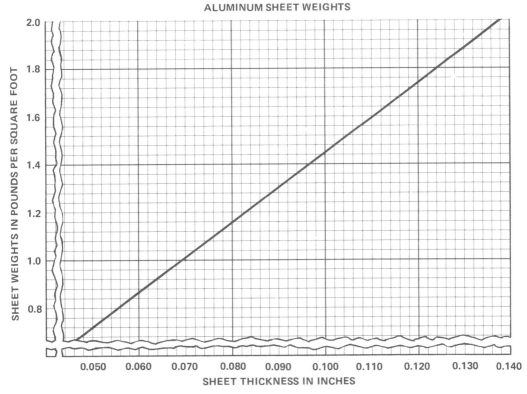

a. What is the weight of each thickness of aluminum sheet listed on the graph? Express the answers to the nearer 0.04 pound per square foot.
b. Find to the nearer 0.04 pound per square foot, the weight of an aluminum sheet 0.056" thick.
c. Find, to the nearer 0.04 pound per square foot, the weight of an aluminum sheet 0.082" thick.
d. Find, to the nearer 0.04 pound per square foot, the weight of an aluminum sheet 0.102" thick.
e. Find, to the nearer 0.04 pound per square foot, the weight of an aluminum sheet 0.128" thick.
f. What thickness sheet of aluminum is required for a sheet 0.8 pound per square foot?
g. What thickness sheet of aluminum is required for a sheet 1.04 pounds per square foot?
h. What thickness sheet of aluminum is required for a sheet 1.36 pounds per square foot?
i. What thickness sheet of aluminum is required for a sheet 1.84 pounds per square foot?

4. Draw a broken-line graph to show the United States' percent of the total world trade for each year of an eight-year period as listed in the table.

YEAR	1st	2nd	3rd	4th	5th	6th	7th	8th
UNITED STATES PERCENT OF TOTAL WORLD TRADE	14.6	14.8	14.6	14.3	13.8	15.4	13.9	13.4

5. Work is determined by multiplying a force by the distance through which the force moves.

$W = f \times d$ where W = work in foot-pounds
f = force in pounds
d = distance in feet

One foot-pound of work is accomplished when a one-pound weight is lifted one foot vertically as shown.

ONE FOOT-POUND OF WORK

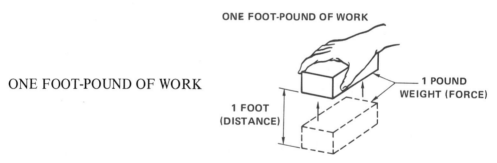

ONE FOOT-POUND OF WORK

1 POUND WEIGHT (FORCE)

1 FOOT (DISTANCE)

Draw a graph showing the relationship of the weight to the work. Use a constant distance value of 15 feet and weight values of 0, 10, 20, 30, 40, and 50 pounds.

Note: It is helpful to make a table of values to be graphed.

6. This formula is used for finding current in an electrical circuit when power and resistance are known.

$I = \sqrt{\dfrac{P}{R}}$ where I = current in amperes
P = power in watts
R = resistance in ohms

Draw a graph using a constant power value of 250 watts. Use resistance values of 25, 50, 75, 100, 125, 150, and 175 ohms.

Note: It is helpful to make a table of the values to be graphed.

7. Weights of round steel bars of different diameters are listed in the table. Copy the table and find the missing changes in weight values. If the table data are plotted, would a straight line or a curved line result? Do not graph the table data.

DIAMETERS IN CENTIMETRES	2.54	2.86	3.18	3.50	3.82	4.14
WEIGHTS IN KILOGRAMS PER METRE OF LENGTH	4.11	5.03	6.21	7.51	8.94	10.49
CHANGES IN WEIGHTS IN KILOGRAMS PER METRE OF LENGTH		0.92	1.18	1.30	1.43	1.55

A curved line would result.

8. The pressure of a liquid increases as the depth of the liquid increases. The table lists water pressures in pounds per square inch at various depths. Copy the table and fill in the missing changes in pressure values. If the table data are plotted, would a straight line or a curved line result? Do not graph the table data.

DEPTH OF WATER IN FEET	4	8	12	16	20	24
WATER PRESSURE IN POUNDS PER SQUARE INCH	1.73	3.46	5.19	6.92	8.65	10.38
CHANGES IN WATER PRESSURE IN POUNDS PER SQUARE INCH		1.73	1.73	1.73	1.73	1.73

A straight line would result.

SECTION 5
Measure

UNIT 25
English Units of Linear Measure

OBJECTIVES

After studying this unit you should be able to

- Express lengths as larger English linear units.
- Express lengths as larger English linear compound numbers.
- Express lengths as smaller English linear units.
- Express lengths as smaller English linear compound numbers.
- Add, subtract, multiply, and divide compound numbers in the English system.

The ability to measure with tools and instruments and to compute with measurements is required in almost all occupations.

Measurement is the comparison of a quantity with a standard unit. A linear measurement is a means of expressing the distance between two points; it is the measurement of lengths. A linear measurement has two parts: a unit of length and a multiplier.

The measurements 2.5 inches and 15¼ miles are examples of denominate numbers. A *denominate number* is a number that refers to a special unit of measure. A *compound denominate number* consists of more than one unit of measure, such as 7 feet 2 inches.

ENGLISH LINEAR UNITS

The yard is the standard unit of linear measure in the English system. From the yard, other units such as the inch and foot are established. The smallest unit is the inch.

ENGLISH UNITS OF LINEAR MEASURE
1 foot (ft) = 12 inches (in)
1 yard (yd) = 3 feet (ft)
1 yard (yd) = 36 inches (in)
1 rod (rd) = 16.5 feet (ft)
1 furlong = 220 yards (yd)
1 mile (mi) = 5 280 feet (ft)
1 mile (mi) = 1 760 yards (yd)
1 mile (mi) = 320 rods (rd)
1 mile (mi) = 8 furlongs

EXPRESSING EQUIVALENT UNITS OF MEASURE

Either of two methods may be used in expressing equivalent units of measure. The two methods apply to all units of measure.

METHOD 1: This is the practical method used for on-the-job applications.
METHOD 2: This method multiplies the given unit of measure by a fraction equal to one.

All of the examples are solved using the first method. In addition, some of the examples are solved using both methods.

EXPRESSING SMALLER ENGLISH UNITS
OF LINEAR MEASURE AS LARGER UNITS

METHOD 1: To express a smaller unit of length as a larger unit of length, divide the given length by the number of smaller units contained in one of the larger units.

Examples:

1. Express 76.5 inches as feet.

   METHOD 1
   Since 12 inches equal 1 foot,
   divide 76.5 by 12.

   $$76.5 \div 12 = 6.375$$
   $$76.5 \text{ inches} = 6.375 \text{ feet } Ans$$

   METHOD 2
   Multiply 76.5 inches by 1
   in the form of $\dfrac{1 \text{ ft}}{12 \text{ in}}$

   $$\dfrac{76.5 \text{ in}}{1} \times \dfrac{1 \text{ ft}}{12 \text{ in}} = \dfrac{76.5 \text{ ft}}{12} = 6.375 \text{ ft } Ans$$

   In the numerator and denominator, divide by the common factor, 1 inch.
   Divide 76.5 ft by 12.

2. How many miles are in 1 320 yards.

   METHOD 1
   Since 1 760 yards equal 1 mile,
   divide 1 320 by 1 760.

   $$1\,320 \div 1\,760 = 0.75$$
   $$1\,320 \text{ yards} = 0.75 \text{ mile } Ans$$

   METHOD 2
   Multiply 1 320 yards
   by 1 in the form of $\dfrac{1 \text{ mi}}{1\,760 \text{ yd}}$

   $$\dfrac{1\,320 \text{ yd}}{1} \times \dfrac{1 \text{ mi}}{1\,760 \text{ yd}} = \dfrac{1\,320 \text{ mi}}{1\,760} = 0.75 \text{ mi } Ans$$

   In the numerator and denominator, divide by the common factor, 1 yard.
   Divide 1 320 mi by 1 760.

Unit 25 English Units of Linear Measure 277

For actual on-the-job applications, smaller units are often expressed as a combination of larger and smaller units (compound denominate numbers).

> Example: A carpenter wants to express $134\frac{7}{8}$ inches as feet and inches.
>
> Since 12 inches equal 1 foot, divide 134 7/8 by 12. There are 11 feet plus a remainder of 2 7/8 inches in 134 7/8 inches.
>
> The carpenter uses 11 feet 2 7/8 inches as an actual on-the-job measurement. *Ans*
>
> $$\begin{array}{r} 11 \phantom{134\tfrac{7}{8}} \\ 12 \overline{)134\tfrac{7}{8}} \\ \underline{12} \\ 14 \\ \underline{12} \\ 2\tfrac{7}{8} \end{array}$$
>
> remainder
>
>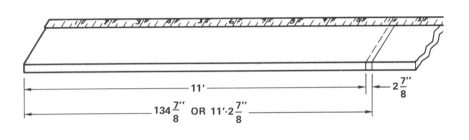

Exercise 25-1

Express each length as indicated.

1. 48 inches as feet
2. 27 inches as feet
3. 21 feet as yards
4. 34 feet as yards
5. 6 336 feet as miles
6. 4 048 inches as yards
7. 44.4 inches as feet
8. 4 928 yards as miles
9. 45.9 feet as yards
10. $53\frac{1}{4}$ feet as yards
11. 200 rods as miles
12. 6 furlongs as miles
13. 75 inches as feet and inches
14. 40 inches as feet and inches
15. 2 420 yards as miles and yards
16. 12 100 feet as miles and feet
17. $127\frac{1}{2}$ inches as feet and inches
18. 63.2 feet as yards and feet
19. $1\,925\frac{1}{3}$ yards as miles and yards
20. $678\frac{3}{4}$ rods as miles and rods

EXPRESSING LARGER ENGLISH UNITS OF LINEAR MEASURE AS SMALLER UNITS

METHOD 1: To express a larger unit of length as a smaller unit of length, multiply the given length by the number of smaller units contained in one of the larger units.

Examples:

1. Express $2\frac{1}{8}$ yards as inches.

   METHOD 1
   Since 36 inches equal 1 yard, multiply 2 1/8 by 36.

   $$2\frac{1}{8} \times 36 = 76\frac{1}{2}$$

   $$2\frac{1}{8} \text{ yards} = 76\frac{1}{2} \text{ inches } Ans$$

   METHOD 2
   Multiply 2 1/8 yards by 1 in the form of $\frac{36 \text{ in}}{1 \text{ yd}}$

   $$\frac{\cancel{17}\,\text{yd}}{\cancel{8}_{\,2}}^{17} \times \frac{\cancel{36}\,\text{in}}{\cancel{1}\,\text{yd}}^{9\,\text{in}} = \frac{153 \text{ in}}{2} = 76\frac{1}{2} \text{ in } Ans$$

   Divide the numerator and denominator by the common factors.

2. How many feet are in 1.62 miles?

   METHOD 1
   Since 5 280 feet equal 1 mile, multiply 1.62 by 5 280.

   $$1.62 \times 5\,280 = 8\,553.6$$
   $$1.62 \text{ miles} = 8\,553.6 \text{ feet } Ans$$

   METHOD 2
   Multiply 1.62 miles by 1 in the form of $\frac{5\,280 \text{ ft}}{1 \text{ mi}}$

   $$\frac{\cancel{1.62\,\text{mi}}}{1} \times \frac{5\,280 \text{ ft}}{\cancel{1\,\text{mi}}} = 8\,553.6 \text{ ft } Ans$$

3. Find the number of inches in 4 feet 3 5/16 inches.

   METHOD 1
   Find the number of inches in 4 feet. $4 \times 12 = 48$

   Add 48 inches and 3 5/16 inches. $48 \text{ inches} + 3\frac{5}{16} \text{ inches} = 51\frac{5}{16} \text{ inches } Ans$

   METHOD 2
   Multiply 4 feet by 1 in the form of $\frac{12 \text{ in}}{1 \text{ ft}}$

   $$\frac{\cancel{4\,\text{ft}}}{1}^{4} \times \frac{12 \text{ in}}{\cancel{1\,\text{ft}}} = 48 \text{ in}$$

   Add 48 inches and 3 5/16 inches. $48 \text{ in} + 3\frac{5}{16} \text{ in} = 51\frac{5}{16} \text{ in } Ans$

In practical applications, larger units are often expressed as two different smaller compound units.

Example: A sheet metal technician wants to express 2 3/4 yards as feet and inches. There is more than one method of solving this problem. One method is shown here.

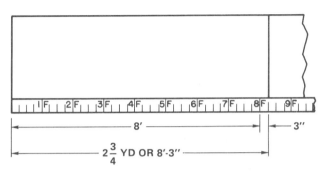

Express $2\frac{3}{4}$ yards as feet.
Since 3 feet equal 1 yard, multiply $2\frac{3}{4}$ by 3. $2\frac{3}{4} \times 3 = 8\frac{1}{4}$.
Therefore, $2\frac{3}{4}$ yards = $8\frac{1}{4}$ feet.
Express $\frac{1}{4}$ foot as inches.
Since 12 inches equal 1 foot, multiply $\frac{1}{4}$ by 12. $\frac{1}{4} \times 12 = 3$.
Therefore, $\frac{1}{4}$ foot = 3 inches.

Combine feet and inches.
$2\frac{3}{4}$ yards = 8 feet plus 3 inches or 8 feet 3 inches.

The sheet metal technician uses a working measurement of 8 feet 3 inches as shown in the figure. *Ans*

Exercise 25-2

Express each length as indicated.

1. 7 feet as inches
2. 2 yards as inches
3. 16 yards as feet
4. $\frac{1}{2}$ mile as yards
5. 1.35 miles as feet
6. $9\frac{1}{6}$ feet as inches
7. 4.25 yards as feet
8. $2\frac{3}{8}$ miles as yards
9. $\frac{1}{8}$ mile as feet
10. $5\frac{1}{10}$ yards as inches

11. 0.38 mile as rods
12. 3.6 miles as furlongs
13. $6\frac{1}{2}$ yards as feet and inches
14. 8.25 yards as feet and inches
15. 0.12 mile as yards and feet
16. $\frac{5}{12}$ mile as yards and feet
17. 2.18 miles as rods and yards
18. $8\frac{7}{32}$ yards as feet and inches
19. 0.9 yard as feet and inches
20. 0.87 mile as yards and feet

ARITHMETIC OPERATIONS WITH COMPOUND NUMBERS

Basic arithmetic operations with compound numbers are often required for on-the-job applications. For example, a material estimator may compute the stock requirements of a certain job by adding 16 feet 7 1/2 inches and 9 feet 10 inches. An iron worker may be required to divide a 14 foot 10 inch beam in 3 equal parts.

The method generally used for occupational problems is given for each basic operation.

ADDITION OF COMPOUND NUMBERS

To add compound numbers, arrange like units in the same column, then add each column. When necessary, simplify the sum.

Examples:

1. Determine the amount of stock, in feet and inches, required to make the welded angle bracket shown.

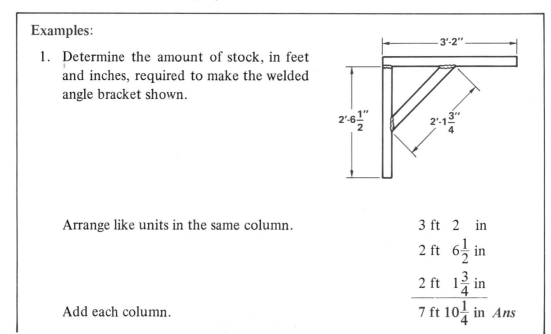

Arrange like units in the same column.

    3 ft 2 in
    2 ft $6\frac{1}{2}$ in
    2 ft $1\frac{3}{4}$ in

Add each column.

    7 ft $10\frac{1}{4}$ in *Ans*

2. Find the sum of the following three lengths of material: 7 yards 2 feet 9 inches, 3 yards 1 foot 10 inches, and 4 yards 2 feet 8 inches.

Arrange like units in the same column.	7 yd 2 ft 9 in
	3 yd 1 ft 10 in
	4 yd 2 ft 8 in
Add each column.	14 yd 5 ft 27 in
Simplify the sum starting at the right.	
Divide 27 by 12 to express 27 inches	14 yd 5 ft
as 2 feet 3 inches.	2 ft 3 in
Add 2 feet to 5 feet.	14 yd 7 ft 3 in
Divide 7 by 3 to express 7 feet as	14 yd 3 in
2 yards 1 foot.	2 yd 1 ft
Add 2 yards to 14 yards.	16 yd 1 ft 3 in *Ans*

Exercise 25-3

Add. Express each answer in the same units as those given in the exercise. Regroup the answer when necessary.

1. 5 ft 6 in + 7 ft 3 in
2. 3 ft 9 in + 4 ft 8 in
3. 6 ft $3\frac{3}{8}$ in + 4 ft $1\frac{1}{2}$ in + 8 ft $10\frac{1}{4}$ in
4. 2 yd 1 ft + 3 yd 1 ft + 6 yd $\frac{1}{2}$ ft
5. 3 yd 2 ft + 5 yd $\frac{1}{4}$ ft + 9 yd $2\frac{3}{4}$ ft
6. 9 yd 2 ft 3 in + 2 yd 0 ft 6 in
7. 12 yd 2 ft 8 in + 10 yd 2 ft 7 in
8. 4 yd 1 ft $3\frac{1}{2}$ in + 7 yd 0 ft 9 in + 4 yd 2 ft 0 in
9. 3 rd 4 yd + 2 rd 1 yd
10. 6 rd $3\frac{1}{4}$ yd + 8 rd 4 yd
11. 1 mi 150 rd + 1 mi 285 rd
12. 3 mi 75 rd 2 yd + 2 mi 150 rd $3\frac{1}{4}$ yd + 1 mi 200 rd 5 yd

SUBTRACTION OF COMPOUND NUMBERS

To subtract compound numbers, arrange like units in the same column, then subtract each column starting from the right. Regroup as necessary.

Examples:

1. Determine length **A** of this pipe.

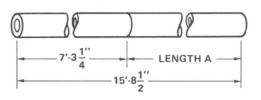

Arrange like units in the same column.

$\qquad$ 15 ft $8\frac{1}{2}$ in

$\qquad$ 7 ft $3\frac{1}{4}$ in

Subtract each column. $\qquad$ 8 ft $5\frac{1}{4}$ in *Ans*

2. Subtract 8 yards 2 feet 7 inches from 12 yards 1 foot 3 inches.

Arrange like units in the same column. $\qquad$ 12 yd 1 ft 3 in

$\qquad$ 8 yd 2 ft 7 in

Subtract each column.

Since 7 inches cannot be subtracted from 3 inches, subtract 1 foot from the foot column (leaving 0 feet). Express 1 foot as 12 inches; then add 12 inches to 3 inches.

12 yd 0 ft 15 in
8 yd 2 ft 7 in

Since 2 feet cannot be subtracted from 0 feet, subtract 1 yard from the yard column (leaving 11 yards). Express 1 yard as 3 feet; then add 3 feet to 0 feet.
Subtract each column.

11 yd 3 ft 15 in
8 yd 2 ft 7 in
3 yd 1 ft 8 in *Ans*

Exercise 25-4

Subtract. Express each answer in the same units as those given in the exercise. Regroup the answer when necessary.

1. 6 ft 7 in − 2 ft 4 in
2. 15 ft 3 in − 12 ft 9 in
3. 10 ft $1\frac{3}{8}$ in − 7 ft $8\frac{7}{16}$ in
4. 8 yd $1\frac{1}{2}$ ft − 4 yd $2\frac{3}{4}$ ft
5. 20 yd 0 ft − 18 yd 2.5 ft
6. 7 yd 1 ft 9 in − 2 yd 2 ft 11 in
7. 16 yd 2 ft 2.15 in − 14 yd 2 ft 4.25 in
8. 23 yd 1 ft 0 in − 3 yd 0 ft $6\frac{5}{8}$ in
9. 5 rd 3 yd 2 ft − 4 rd 2 yd 1 ft
10. 2 rd 5 yd $1\frac{1}{3}$ ft − 1 rd 0 yd $1\frac{2}{3}$ ft
11. 7 mi 240 rd − 3 mi 310 rd
12. 4 mi 150 rd 4 yd − 1 mi 175 rd 5 yd

MULTIPLICATION OF COMPOUND NUMBERS

To multiply compound numbers, multiply each unit of the compound number by the multiplier. When necessary, simplify the product.

Examples:

1. A plumber cuts 4 pieces of copper tubing. Each piece is 8 feet $2\frac{3}{4}$ inches long. Determine the total length of tubing required.

    Multiply each unit by 4.

    $$\begin{array}{r} 8 \text{ ft } 2\frac{3}{4} \text{ in} \\ 4 \\ \hline 32 \text{ ft } 11 \text{ in } Ans \end{array}$$

2. Multiply 12 yards 2 feet 9 inches by 7.5.

    Multiply each unit by 7.5.

    $$\begin{array}{r} 12 \text{ yd } 2 \text{ ft } 9 \text{ in} \\ 7.5 \\ \hline 90 \text{ yd } 15 \text{ ft } 67.5 \text{ in} \end{array}$$

    Simplify the product starting at the right.

    Divide 67.5 by 12 to express 67.5 inches as 5 feet 7.5 inches. Add 5 feet to 15 feet.

    $$\begin{array}{r} 90 \text{ yd } 15 \text{ ft} \\ 5 \text{ ft } 7.5 \text{ in} \\ \hline 90 \text{ yd } 20 \text{ ft } 7.5 \text{ in} \end{array}$$

    Divide 20 by 3 to express 20 feet as 6 yards 2 feet.

    $$\begin{array}{r} 90 \text{ yd } \phantom{0 \text{ ft }} 7.5 \text{ in} \\ 6 \text{ yd } 2 \text{ ft} \\ \hline 96 \text{ yd } 2 \text{ ft } 7.5 \text{ in } Ans \end{array}$$

Exercise 25-5

Multiply. Express each answer in the same units as those given in the exercise. Regroup the answer when necessary.

1. 7 ft 3 in × 2
2. 4 ft 5 in × 3
3. 12 ft 3 in × 5.5
4. 6 yd $\frac{1}{2}$ ft × 4
5. 18 yd 2 ft × 7
6. 5 yd 1.25 ft × 4.8
7. 9 yd 2 ft 3 in × 2
8. 11 yd 1 ft $7\frac{3}{4}$ in × 3
9. 10 yd 2 ft 9 in × $\frac{1}{2}$
10. 6 rd 4 yd × 5
11. 5 mi 210 rd × 1.4
12. 3 mi 180 rd 5 yd × 2

DIVISION OF COMPOUND NUMBERS

To divide compound numbers, divide each unit by the divisor starting at the left. If a unit is not exactly divisible, express the remainder as the next smaller unit and add it to the given number of smaller units. Continue the process until all units are divided.

Examples:

1. The 4 holes in this I beam are equally spaced. Determine the distance between 2 consecutive holes.

   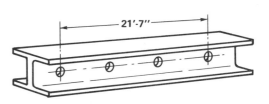

   Since there are 3 spaces between holes, divide 21 feet 7 inches by 3.

   $$3 \overline{\smash{)}\, 21 \text{ ft } 7 \text{ in}}^{\,7 \text{ ft } 2\frac{1}{3} \text{ in } Ans}$$

2. Divide 28 yards 1 foot 4 inches by 8.

Divide 28 yards by 8.	28 yd ÷ 8 = 3 yd (quotient) and a 4 yd remainder.
Express the 4 yard remainder as 12 feet. Add 12 feet to the 1 foot given in the problem.	4 yd = 4 × 3 ft = 12 ft 12 ft + 1 ft = 13 ft
Divide 13 feet by 8.	13 ft ÷ 8 = 1 ft (quotient) and a 5 ft remainder.
Express the 5 foot remainder as 60 inches.	5 × 12 in = 60 in
Add 60 inches to the 4 inches given in the problem.	60 in + 4 in = 64 in
Divide 64 inches by 8.	64 in ÷ 8 = 8 in
Collect quotients.	3 yd 1 ft 8 in *Ans*

Exercise 25-6

Divide. Express each answer in the same units as those given in the exercise.

1. 9 ft 6 in ÷ 3
2. 5 ft 2 in ÷ 2
3. 18 ft 3.9 in ÷ 4
4. 16 yd 2 ft ÷ 8

5. 21 yd 1 ft ÷ $1\frac{1}{2}$

6. 4 yd 3.75 ft ÷ 3

7. 14 yd 2 ft 6 in ÷ 2

8. 17 yd 1 ft 10 in ÷ 5

9. 6 yd 2 ft $3\frac{1}{4}$ in ÷ 0.5

10. 5 rd 2 yd ÷ 4

11. 3 mi 150 rd ÷ $1\frac{1}{2}$

12. 4 mi 310 rd 4 yd ÷ 3

UNIT REVIEW

Exercise 25-7

Express each length as indicated.

1. 75 inches as feet
2. $40\frac{1}{2}$ inches as feet
3. 17 feet as yards
4. 22.5 feet as yards
5. 558 inches as yards
6. 6 336 feet as miles
7. $13\frac{3}{4}$ yards as rods
8. 160 rods as miles
9. 87 inches as feet and inches
10. 52 feet as yards and feet
11. 3 050 yards as miles and yards
12. 10 700 feet as miles and feet
13. $17\frac{3}{4}$ yards as rods and yards
14. 472 rods as miles and rods
15. 9 feet as inches
16. 0.25 feet as inches
17. 23 yards as feet
18. $\frac{3}{4}$ mile as yards
19. 0.3 mile as feet
20. $2\frac{1}{8}$ yards as inches
21. 3 rods as yards
22. 1.7 miles as rods
23. $4\frac{3}{4}$ yards as feet and inches
24. 0.64 mile as yards and feet
25. $\frac{1}{12}$ mile as yards and feet
26. 7.3 yards as feet and inches
27. 3 rods as yards and feet
28. 0.76 mile as rods and yards

Add. Express each answer in the same units as those given in the exercise. Regroup the answer when necessary.

29. 6 ft 8 in + 7 ft 2 in
30. 5 ft 9 in + 8 ft $3\frac{3}{4}$ in
31. 2 yd 2 ft + 1 yd 2 ft
32. 3 ft $7\frac{1}{2}$ in + 5 ft 11 in + 4 ft $10\frac{1}{4}$ in

33. 7 yd 2 ft + 10 yd $\frac{2}{3}$ ft + 9 yd 1$\frac{3}{4}$ ft

34. 2 rd 4 yd + 3 rd 3$\frac{1}{4}$ yd + 1 rd $\frac{1}{2}$ yd

35. 1 mi 210 rd + 2 mi 50 rd + 1 mi 300 rd

36. 2 mi 100 rd 4 yd + 1 mi 280 rd 3.6 yd

Subtract. Express each answer in the same units as those given in the exercise. Regroup the answer when necessary.

37. 14 ft 8 in − 10 ft 7 in

38. 19 ft $\frac{3}{4}$ in − 8 ft 11 in

39. 13 yd 2 ft − 6 yd 1$\frac{1}{2}$ ft

40. 23 yd $\frac{1}{2}$ ft − 2 yd 2 ft

41. 20 yd 2 ft 8 in − 18 yd 1 ft 3$\frac{1}{8}$ in

42. 11 yd 1 ft 1.25 in − 4 yd 2 ft 7.5 in

43. 3 rd 4 yd 2 ft − 2 rd 3 yd 2$\frac{1}{3}$ ft

44. 3 mi 125 rd 3 yd − 1 mi 175 rd 5 yd

Multiply. Express each answer in the same units as those given in the exercise. Regroup the answer when necessary.

45. 3 ft 2 in × 4

46. 2 ft 8$\frac{1}{2}$ in × 3

47. 4 yd 2 ft × 2

48. 2 yd 2 ft 7 in × 5

49. 6 yd 2 ft 3 in × 6$\frac{1}{2}$

50. 8 rd 3 yd × 6

51. 2 mi 175 rd × 2.5

52. 2 mi 120 rd 3$\frac{1}{3}$ yd × 3

Divide. Express each answer in the same units as those given in the exercise.

53. 15 ft 10 in ÷ 5

54. 9 ft 3 in ÷ 2

55. 18 yd 2 ft ÷ 3

56. 20 yd $\frac{1}{2}$ ft ÷ 2$\frac{1}{2}$

57. 10 yd 2 ft 4 in ÷ 2

58. 10 yd 1 ft 8 in ÷ 4

59. 6 rd 4 yd ÷ 5

60. 5 mi 250 rd 5 yd ÷ 4

Unit 25 English Units of Linear Measure   287

PRACTICAL APPLICATIONS

Exercise 25-8

1. The first-floor plan of a house is shown. Find distances **A, B, C,** and **D** in feet and inches.

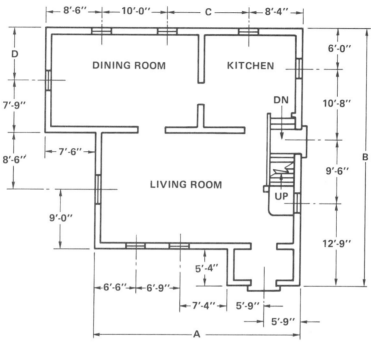

2. A surveyor subdivides a parcel of land in 5 lots of equal width as shown. Find the number of feet in distances **A** and **B**.

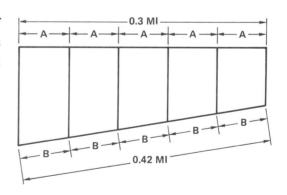

3. A bolt (roll) contains 70 yards 2 feet of fabric. The following lengths of fabric are sold from the bolt: 4 yards 2 feet, 5 yards 1 1/4 feet, 7 yards 2 1/2 feet. Find the length of fabric left on the bolt. Express the answer in yards and feet.

4. A building construction assistant lays out this stairway.
   a. Find, in feet and inches, the total run of the stairs.
   b. Find, in feet and inches, the total rise of the stairs.

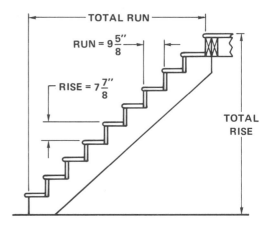

5. A structural steel fabricator cuts 4 equal lengths from a channel iron. Allow 1/8" waste for each cut. Find the length, in feet and inches, of the remaining piece.

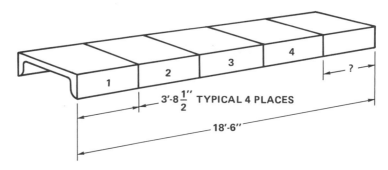

6. The floor of a room is to be covered with oak flooring. The flooring is 2 1/4 inches wide. Allow 320 linear feet for waste. Find the total number of linear feet of oak flooring, including waste, needed for the floor.

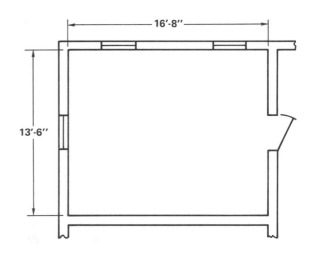

7. An apparel maker must know the fabric cost per garment. Find the material cost of a garment which requires 68 inches of fabric at $4.75 per yard and 52 inches of lining at $2.20 per yard.

8. A concrete beam is 12 1/3 yards long. Find distances **A**, **B**, and **C** in feet and inches.

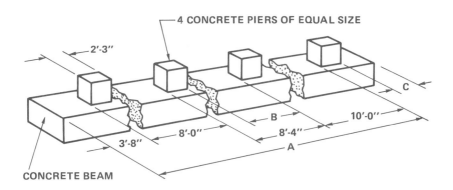

9. A carpet installer contracts to supply and install carpeting in the hallways of an office building. The locations of the hallways are shown. The hallways are 4 1/2 feet wide. The price charged for both carpet cost and installation is $18.75 per running (linear) yard. Find the total cost of the installation job.

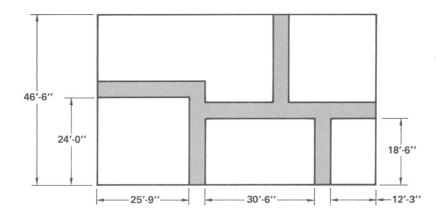

10. The table frame shown is to be made of welded square tubing. To find the total weight of the frame, a welder first looks up the weight of one foot of the size tubing to be used. The weight is 3.6 pounds per foot of length. Find the total weight to the nearer tenth pound.

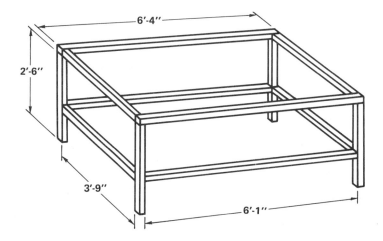

UNIT 26
Metric Units of Linear Measure

OBJECTIVES

After studying this unit you should be able to
- Express metric linear measurements as larger or smaller metric linear measurements.
- Perform arithmetic operations with metric linear units.
- Express metric length units as English length units.
- Solve practical applied metric length problems.
- Solve practical applied linear metric-English equivalent problems.

The seven basic units of measure used in the metric system are shown in this table.

SEVEN BASIC UNITS IN THE METRIC SYSTEM		
QUANTITY	BASIC UNIT	SYMBOL
Length	metre	m
Mass	kilogram	kg
Time	second	s
Electric current	ampere	A
Temperature	kelvin	K
Light intensity	candela	cd
Amount of substance	mole	mol

The following table lists the most frequently used metric units.

KM HM DKM M DM CM MM

FREQUENTLY USED METRIC UNITS		
QUANTITY	UNIT	SYMBOL
length	kilometre	km
	metre	m
	centimetre	cm
	millimetre	mm
area	square kilometre	km^2
	hectare (10 000 m²)	ha
	square metre	m^2
	square centimetre	cm^2
volume or capacity	cubic metre	m^3
	cubic decimetre	dm^3
	litre	L
	cubic centimetre	cm^3
	millilitre	mL
mass (weight)	metric ton (1 000 kg)	t
	kilogram	kg
	gram	g
	milligram	mg
time	day	d
	hour	h
	minute	min
	second	s
temperature	degree Celsius	°C
speed or velocity	metre per second	m/s
	kilometre per hour	km/h
plane angle	degree	°
force	kilonewton	kN
	newton	N
pressure	kilopascal	kPa
acceleration	metre per second squared	m/s^2
rotational frequency	revolution per second	r/s
	revolution per minute	r/min
density	kilogram per cubic metre	kg/m^3
	gram per litre	g/L

Not all of the units listed in the tables are used in this book. Some of the units are applied only in very specific uses. Metric units are discussed in this book as they apply to the particular material presented. Only linear (length) metric measurement units are used here.

ADVANTAGES OF USING METRICS

An advantage of the metric system is that it allows easy, fast computations. Since metric system units are based on powers of ten, figuring is simplified. To express a certain metric unit as a larger or smaller metric unit, all that is required is to move the decimal point a proper number of places to the left or right. Metric system units are also easy to learn.

The metric system does not require difficult conversions as in the English system. It is easier to remember that 1 000 metres equal 1 kilometre than to remember 1 760 yards equal 1 mile.

METRIC UNITS OF LINEAR MEASURE

The following metric power of ten prefixes are based on the metre:

milli means one thousandth (0.001)
centi means one hundredth (0.01)
deci means one tenth (0.1)

deka means ten (10)
hecto means hundred (100)
kilo means thousand (1 000)

The following table lists the metric units of length with their symbols. These units are based on the metre. Observe that each unit is ten times greater than the unit directly above it.

METRIC UNITS OF LINEAR MEASURE			
1 millimetre (mm)	= 0.001 metre (m)	1 000 millimetres (mm)	= 1 metre (m)
1 centimetre (cm)	= 0.01 metre (m)	100 centimetres (cm)	= 1 metre (m)
1 decimetre (dm)	= 0.1 metre (m)	10 decimetres (dm)	= 1 metre (m)
1 metre (m)	= 1 metre (m)	1 metre (m)	= 1 metre (m)
1 dekametre (dam)	= 10 metres (m)	0.1 dekametre (dam)	= 1 metre (m)
1 hectometre (hm)	= 100 metres (m)	0.01 hectometre (hm)	= 1 metre (m)
1 kilometre (km)	= 1 000 metres (m)	0.001 kilometres (km)	= 1 metre (m)

The most frequently used metric units of length are the kilometre (km), metre (m), centimetre (cm), and millimetre (mm). In actual applications, the dekametre (dam) and hectometre (hm) are not used. The decimetre (dm) is seldom used.

RULES FOR WRITING METRIC QUANTITIES

Notice that periods are not used after the unit symbols listed in the table. For example, write 1mm, *not* 1 m.m. or 1mm., when expressing the millimetre unit as a symbol.

A comma is *not* used to separate digits in groups of three. Many countries use commas for decimal markers. To avoid confusion, a space is left between groups of three digits counting from the decimal point.

Examples:

Write 1 070 m, *not* 1,070 m
Write 2 753 000 km, *not* 2,753,000 km
Write 0.656 25 cm, *not* 0.65625 cm

EXPRESSING EQUIVALENT UNITS WITHIN THE METRIC SYSTEM

To express a given unit of length as a larger unit, move the decimal point a certain number of places to the left. To express a given unit of length as a smaller unit, move the decimal point a certain number of places to the right. The exact procedure of moving decimal points is shown in the following examples. Refer to the metric units of linear measure table.

Examples:
1. Express 65 decimetres (dm) as metres (m).

   Since a metre is the next larger unit to a decimetre, move the decimal point 1 place to the left.   6.5.

   In moving the decimal point 1 place to the left, you are actually dividing by 10.

   65 dm = 6.5 m *Ans*

2. Express 0.28 decimetre (dm) as centimetres (cm).

   Since a centimetre is the next smaller unit to a decimetre, move the decimal point 1 place to the right.   0.2.8

   In moving the decimal point 1 place to the right, you are actually multiplying by 10.

   0.28 dm = 2.8 cm *Ans*

3. Express 0.378 metre (m) as millimetres (mm).

   Expressing metres as millimetres involves 3 steps.

   0.378 m   =   3.78 dm   =   37.8 cm   =   378 mm
        ①   ②   ③

   Since a millimetre is three smaller units from a metre, move the decimal point 3 places to the right. In moving the decimal point 3 places to the right, you are actually multiplying by 10^3 (10 × 10 × 10), or 1 000.

   0.378.

   0.378 m = 378 mm *Ans*

Exercise 26-1

Express these lengths in metres.
1. 34 decimetres
2. 4 320 millimetres
3. 0.05 kilometres
4. 2.58 dekametres
5. 240 millimetres
6. 95.6 centimetres
7. 0.84 hectometres
8. 402 decimetres
9. 1.05 kilometres
10. 56.9 millimetres
11. 14.8 dekametres
12. 2 070 centimetres

Express each value as indicated.
13. 7 decimetres as centimetres
14. 28 millimetres as centimetres
15. 5 centimetres as millimetres
16. 0.38 metre as dekametres
17. 2.4 kilometres as hectometres
18. 27 dekametres as metres
19. 310.6 decimetres as metres
20. 3.9 hectometres as kilometres
21. 735 millimetres as decimetres
22. 12 metres as centimetres
23. 616 metres as kilometres
24. 404 dekametres as decimetres
25. 0.08 kilometres as decimetres
26. 8 975 millimetres as dekametres
27. 0.06 hectometres as centimetres
28. 206 decimetres as kilometres

ARITHMETIC OPERATIONS WITH METRIC LENGTHS

Arithmetic operations are performed with metric denominate numbers the same as with English denominate numbers. Compute the arithmetic operations, then write the proper metric unit of measure.

Examples:
1. 3.2 m + 5.3 m = **8.5 m**
2. 20.65 mm − 16.32 mm = **4.33 mm**
3. 7.2 × 10.6 cm = **76.32 cm**
4. 24.84 km ÷ 4 = **6.21 km**

As with the English system, only like units can be added or subtracted.

Examples: Add 0.05 metre (m), 6.8 decimetres (dm), 150 centimetres (cm), and 225 millimetres (mm). Express the answer in centimetres (cm).

There is more than one way to solve this problem. The following procedure is the most efficient method.

Express all quantities as centimetres.

0.05 m = 5 cm (decimal point moved 2 places to right)
6.8 dm = 68 cm (decimal point moved 1 place to right)
150 cm = 150 cm
225 mm = 22.5 cm (decimal point moved 1 place to left)

Add.   5 cm + 68 cm + 150 cm + 22.5 cm = **245.5 cm** *Ans*

Exercise 26-2 Every 3

Perform indicated operations. Express the answer in the unit indicated.

1. 3.5 m + 6.1 m = ? m
2. 13.68 cm − 9.32 cm = ? cm
3. 15 × 8.06 mm = ? mm
4. 18.33 km ÷ 3 = ? km
5. 20.7 mm + 3.8 cm + 0.09 cm = ? cm
6. 17.8 hm − 0.9 km = ? km
7. 1.73 km − 850 m = ? m
8. 250 mm + 7.3 dm − 29.4 cm = ? mm
9. 0.02 km + 387.6 m + 37.05 hm = ? m
10. 332.5 cm − 3.3 m = ? cm
11. 67 dam − 0.66 km = ? m
12. 94.3 dm + 872 mm + 8.7 m = ? dm
13. 2.88 hm − 173 m = ? km
14. 40.4 cm + 0.08 m + 2.3 dm + 73.2 mm = ? cm

METRIC-ENGLISH LINEAR EQUIVALENTS

Since both the English and metric systems are used in this country, it is sometimes necessary to express equivalent measurements between systems.

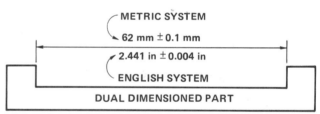

NOTE: ±0.1 mm AND ±0.004 in ARE CALLED TOLERANCES.

For example, some manufacturing industries use dual dimensioning on engineering drawings. Dimensions are shown as both English and metric units. The illustration shows one method of dual dimensioning.

Dual dimensioning permits the manufacturing of many parts in shops using either the English system decimal inch or metric system of measurement. It is a practical method for industries which have plants in foreign countries.

The commonly used equivalent factors of linear measure are listed in the following table.

LINEAR MEASURE			
METRIC TO ENGLISH UNITS		ENGLISH TO METRIC UNITS	
1 millimetre (mm) =	0.039 37 inch (in)	1 inch (in) =	25.4 millimetres (mm)
1 centimetre (cm) =	0.393 7 inch (in)	1 inch (in) =	2.54 centimetres (cm)
1 metre (m) =	39.37 inches (in)	1 foot (ft) =	0.304 8 metre (m)
1 metre (m) =	3.280 8 feet (ft)	1 yard (yd) =	0.914 4 metre (m)
1 kilometre (km) =	0.621 4 mile (mi)	1 mile (mi) =	1.609 kilometres (km)

The relationship between English decimal inch units and metric centimetre units is shown by comparing scales.

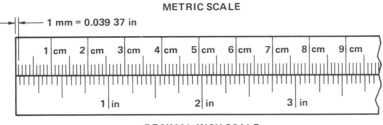

METHOD 1: To express a unit given in one measurement system as a unit in the other system, multiply the given measurement by the appropriate equivalent factor listed in the linear measure table.

Examples:

1. Express 5.2 centimetres (cm) as inches (in). Express the answer to 3 decimal places.

   METHOD 1
   Since 1 cm = 0.393 7 in, 5.2 cm = 5.2 × 0.393 7 in = 2.047 in *Ans*

   METHOD 2
   $$\frac{5.2 \text{ cm}}{1} \times \frac{0.393\ 7 \text{ in}}{1 \text{ cm}} = 2.047 \text{ in } Ans$$

2. Express 3.5 inches (in) as centimetres (cm).

   METHOD 1
   Since 1 in = 2.54 cm, 3.5 in = 3.5 × 2.54 cm = 8.89 cm *Ans*

   METHOD 2
   $$\frac{3.5 \text{ in}}{1} \times \frac{2.54 \text{ cm}}{1 \text{ in}} = 8.89 \text{ cm } Ans$$

298 Section 5 Measure

For actual on-the-job applications, units of measure are often expressed in either the English System or the metric system of measure.

Examples:

1. Determine the total length of the template in millimetres.

   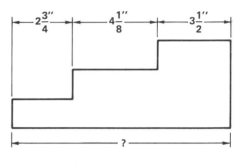

   Do not express each of the mixed numbers as millimetres.

   Add the dimensions as they are given.

   $2\frac{3}{4}$ in + $4\frac{1}{8}$ in + $3\frac{1}{2}$ in = $10\frac{3}{8}$ in

   Express $10\frac{3}{8}$ inches as a mixed decimal.

   $10\frac{3}{8}$ in = 10.375 in

   Express 10.375 inches as millimetres. Since 1 in = 25.4 mm, 10.375 in = 10.375 × 25.4 mm = **263.525 mm** *Ans*

2. A pipe is 11 feet 3 3/4 inches long. Express this length in metres.

   Express 11 feet as inches and add 3 3/4 inches.

   11 ft = 11 × 12 in = 132 in; 132 in + $3\frac{3}{4}$ in = $135\frac{3}{4}$ in

   Express $135\frac{3}{4}$ inches as a mixed decimal. $135\frac{3}{4}$ in = 135.75 in

   Express 135.75 inches as centimetres. Since 1 in = 2.54 cm, 135.75 in = 135.75 × 2.54 cm = 344.805 cm

   Express 344.805 centimetres as metres. Since 1 cm = 0.01 m, move the decimal point 2 places to the left.

   344.805 cm = **3.448 05 m** *Ans*

Practice

1. Refer to this diagram and compute distance **A** in feet. Round the answer to 2 decimal places.

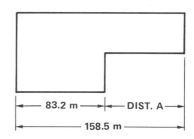

247.04 ft *Ans*

2. Express 1.4 kilometres (km) as feet (ft). Round the answer to the nearer foot.

4 593 ft *Ans*

Exercise 26-3 Friday 3

Express each length as indicated. When necessary, round the answer to 3 decimal places.

1. 12 millimetres as inches
2. 0.3 metre as inches
3. 35 centimetres as inches
4. 4 metres as feet
5. 6 metres as yards
6. 9 kilometres as miles
7. 0.66 metre as feet
8. 0.04 metre as inches
9. 63.3 millimetres as inches
10. 200 millimetres as feet
11. 8.9 centimetres as inches
12. 78 centimetres as feet
13. 112 kilometres as miles
14. 3 millimetres as inches
15. 0.08 kilometre as feet
16. 2 metres as yards
17. 83 centimetres as feet
18. 15.6 metres as yards

Express each length as indicated. When necessary, round the answer to 2 decimal places.

19. 5 inches as millimetres
20. 11 inches as centimetres
21. 1.5 feet as metres
22. 0.25 foot as metres
23. 0.07 inch as millimetres
24. 2.6 miles as kilometres
25. 0.75 mile as kilometres
26. $6\frac{1}{4}$ yards as metres
27. 7.8 inches as centimetres
28. 3 feet as centimetres
29. 50 inches as metres
30. 970 yards as kilometres
31. 3 870 feet as kilometres
32. 16 feet 3 inches as metres
33. $3\frac{1}{4}$ inches as centimetres
34. 8 feet $1\frac{1}{8}$ inches as metres
35. $\frac{3}{4}$ inch as millimetres
36. $235\frac{1}{2}$ yards as kilometres

UNIT REVIEW

Exercise 26-4

Express each value as indicated.

1. 32 millimetres as centimetres
2. 6 centimetres as millimetres
3. 3 480 millimetres as metres
4. 0.49 metre as dekametres
5. 4.6 hectometres as kilometres
6. 18 metres as centimetres
7. 825 metres as kilometres
8. 372 dekametres as decimetres
9. 0.07 kilometre as decimetre
10. 315 decimetres as kilometres
11. 6 510 millimetres as dekametres
12. 0.03 hectometre as centimetres

Perform indicated operations. Express the answer in the unit indicated.

13. 8.6 cm + 18.9 mm = ? mm
14. 2.3 m − 183 cm = ? cm
15. 20 × 7.3 dm = ? dm
16. 12.08 hm ÷ 4 = ? hm
17. 0.05 km + 223.7 m + 26.05 hm = ? m
18. 811.3 cm − 4.2 m = ? cm
19. 83.6 dm + 263 mm + 3.3 m = ? dm
20. 38.8 cm + 3.2 dm + 68.6 mm + 0.02 m = ? cm

Express each length as indicated. When necessary, round the answers to 3 decimal places.

21. 0.6 metre as inches
22. 14 millimetres as inches
23. 3 metres as feet
24. 16 kilometres as miles
25. 0.88 metre as feet
26. 0.09 metre as inches
27. 275 millimetres as feet
28. 82 centimetres as feet
29. 307 kilometres as miles
30. 0.13 kilometres as feet
31. 3.5 metres as yards
32. 74 centimetres as feet

Express each length as indicated. When necessary, round the answers to 2 decimal places.

33. 16 inches as centimetres
34. 7 inches as millimetres
35. 3.6 feet as metres
36. 1.9 miles as kilometres
37. 4 yards as metres
38. 2 feet as centimetres
39. 62 inches as metres
40. 4 932 feet as kilometres
41. 12 feet 8 inches as metres
42. $\frac{7}{8}$ inch as millimetres
43. $782\frac{1}{2}$ yards as kilometres
44. 7 feet $3\frac{3}{4}$ inches as metres

PRACTICE APPLICATIONS

Exercise 26-5

1. Find, in millimetres, dimensions **A, B, C,** and **D** of the plate shown. Round the answers to 2 decimal places.

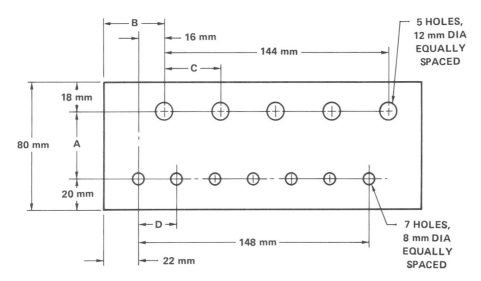

2. Three pieces of stock measuring 3.2 decimetres, 9 centimetres, and 7 centimetres in length are cut from a piece of fabric 0.6 metre long. How many centimetres long is the remaining piece?

3. Find, in metres, the total length of the wall section shown.

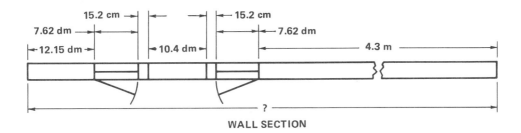

WALL SECTION

4. A car travels round trip from Manchester to New London on the route shown.

a. On the trip to New London the car averages 64 kilometres per hour. How long does the trip to New London take?
b. The return trip from New London to Manchester takes 1.25 hours. How many kilometres per hour does the car average returning to Manchester? Express the answer to the nearer kilometre per hour.

5. The part shown is manufactured in a machine shop using decimal inch machinery and tools. In order to make this part, a drafter expresses millimetre dimensions as decimal inch dimensions. Express dimensions **A–H** as decimal inches to 3 decimal places.

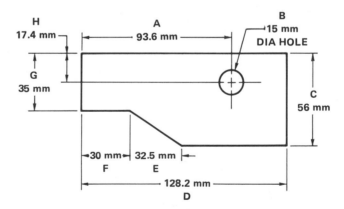

6. Preshrunk fabric is shrunk by the manufacturer. A length of fabric measures 150 metres before shrinking. If the shrinkage is 7 millimetres per metre of length, what is the total length of fabric after shrinking?

7. A thickness (feeler) gauge is shown. This gauge has 9 leaves which range from 0.04 mm to 0.30 mm in thickness. Find the smallest combination of gauge leaves which total each thickness listed. More than one combination may total certain thicknesses.

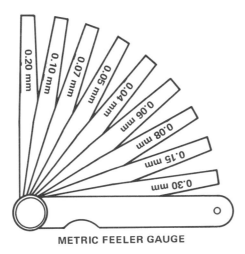

METRIC FEELER GAUGE

a. 0.14 mm
b. 0.13 mm
c. 0.22 mm

d. 0.29 mm
e. 0.33 mm
f. 0.42 mm

g. 0.53 mm
h. 0.69 mm
i. 0.84 mm

8. Find dimensions **A** and **B**, in centimetres, of this pattern.

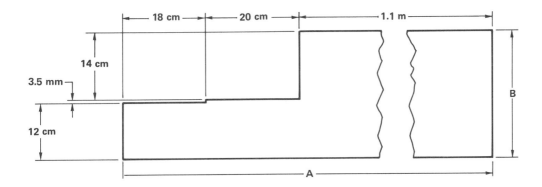

9. Express each velocity in metres per minute. Round answers to 1 decimal place when necessary.
   a. 35 kilometres per hour
   b. 50.8 kilometres per hour
   c. 46.4 kilometres per hour
   d. 60.07 kilometres per hour

304   Section 5   Measure

10. A firm is adopting the metric system in the manufacture of its products. The dimensions on the front view of this lockplate must be expressed as millimetres. Express dimensions **A–G** in millimetres to 2 decimal places.

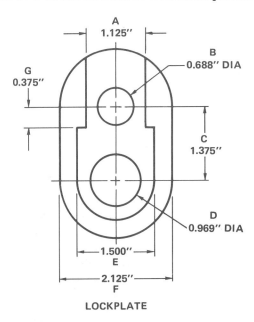

LOCKPLATE

11. For a certain engine overhaul an automobile mechanic grinds the cylinder walls. The diameter of each cylinder before grinding is 95 millimetres. Grinding increases the diameter of each cylinder by 0.76 millimetre. A clearance of 0.065 millimetre is required between the piston and cylinder wall. What size (diameter in millimetres) pistons should be installed in the engine?

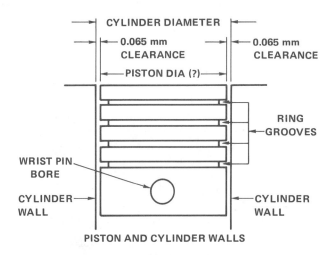

PISTON AND CYLINDER WALLS

12. The room lengths and widths in the floor plan shown are expressed in feet and inches.

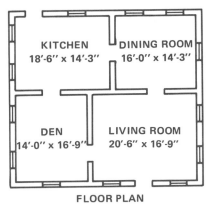

FLOOR PLAN

a. Express, to the nearer thousandth metre, the length and width of the kitchen.
b. Express, to the nearer thousandth metre, the length and width of the dining room.
c. Express, to the nearer thousandth metre, the length and width of the den.
d. Express, to the nearer thousandth metre, the length and width of the living room.

UNIT 27
Equivalent Units of Area Measure

OBJECTIVES

After studying this unit you should be able to

- Express given English area measures as larger and smaller units.
- Express given metric area measures as larger and smaller units.
- Express area measures as English and metric units.
- Solve applied area problems using English and metric units of surface measure.

The ability to compute areas is necessary in many occupations. In agriculture, crop yields and production are determined in relation to land area. Fertilizers and other chemical requirements are computed in terms of land area. In the construction field, carpenters are regularly involved with surface measure, such as floor and roof areas.

ENGLISH UNITS OF SURFACE MEASURE (AREA)

A surface is measured by determining the number of surface units contained in it. A surface is two dimensional. It has length and width, but no thickness. Both length and width must be expressed in the same unit of measure. Area is computed as the product of two linear measures and is expressed in square units. For example, 2 inches $\times$ 4 inches = 8 square inches.

The surface enclosed by a square which is one foot on a side is one square foot. The surface enclosed by a square which is one inch on a side is one square inch. Similar meanings are attached to square yard, square rod, and square mile. For our uses, the statement "area of the surface enclosed by a figure" is shortened to "area of a figure." Therefore, areas will be referred to as the area of a rectangle, area of a triangle, area of a circle, etc.

Look at the reduced size drawing showing a square inch and a square foot. Observe that one linear foot equals 12 linear inches, but one square foot equals 12 inches times 12 inches or 144 square inches.

This table lists common units of surface measure. Other than the unit acre, surface measure units are the same as linear measure units with the addition of the term square.

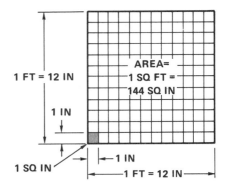

ENGLISH UNITS OF AREA MEASURE	
1 square foot (sq ft)	= 144 square inches (sq in)
1 square yard (sq yd)	= 9 square feet (sq ft)
1 square rod (sq rd)	= 30.25 square yards (sq yd)
1 acre (A)	= 160 square rods (sq rd)
1 acre (A)	= 43 560 square feet (sq ft)
1 square mile (sq mi)	= 640 acres (A)

EXPRESSING ENGLISH AREA MEASURE EQUIVALENTS

METHOD 1: To express a given English unit of area as a larger English unit of area, divide the given area by the number of square units contained in one of the larger units.

Example: Express 720 square inches as square feet.

METHOD 1
Since 144 sq in = 1 sq ft, divide 720 by 144.
720 ÷ 144 = 5; 720 square inches = **5 square feet** *Ans*

METHOD 2

$$\frac{\cancel{720} \text{ sq in}}{1} \times \frac{1 \text{ sq ft}}{\cancel{144} \text{ sq in}} = 5 \text{ sq ft } Ans$$

with 720/1 reduced to 5 and 144 reduced to 1.

Practical Application Examples:

1. A building lot contains 32 670 square feet. How large is the lot in acres?

    Since 43 560 square feet equal 1 acre, divide 32 670 by 43 560.

    $$32\,670 \div 43\,560 = 0.75$$
    $$32\,670 \text{ square feet} = 0.75 \text{ acre } Ans$$

2. How many square yards are there in a drape which contains 3 240 square inches?

    Since 144 square inches equal 1 square foot and 9 square feet equal 1 square yard, there are 9 × 144 or 1 296 square inches in one square yard. Divide 3 240 by 1 296.

    $$3\,240 \div 1\,296 = 2.5$$
    $$3\,240 \text{ square inches} = 2.5 \text{ square yards } Ans$$

METHOD 1: To express a given English unit of area as a smaller English unit of area, multiply the given area by the number of square units contained in one of the larger units.

Example: Express 3.2 square yards as square feet.

METHOD 1
Since 9 sq ft = 1 sq yd, multiply 3.2 by 9.
3.2 × 9 = 28.8; 3.2 square yards = 28.8 square feet *Ans*

METHOD 2

$$\frac{3.2 \, \cancel{\text{sq yd}}}{1} \times \frac{9 \text{ sq ft}}{1 \, \cancel{\text{sq yd}}} = 28.8 \text{ sq ft } Ans$$

Practical Application Examples:

1. A sheet of aluminum which contains 18 square feet is sheared into 40 strips of equal size. What is the area of each strip in square inches?

    Since 144 square inches equal 1 square foot, multiply 144 by 18.

    $$144 \times 18 = 2\,592$$

    18 square feet = 2 592 square inches

    Divide 2 592 square inches by the number of strips (40).

    $$2\,592 \div 40 = 64.8$$

    The area of each strip is 64.8 square inches. *Ans*

2. A land developer purchased 0.2 square mile of land. The land was subdivided into 256 building lots of approximately the same area. What is the average number of square feet per building lot?

    Since 640 acres equal 1 square mile, there are 0.2 × 640 or 128 acres of land.

    The average number of acres per building lot equals the total number of acres (128) divided by the number of lots (256).

    $$128 \div 256 = 0.5 \text{ acre}$$

    Since 43 560 square feet equal 1 acre, multiply 0.5 × 43 560.

    $$0.5 \times 43\,560 = 21\,780$$

    The average area per building lot is 21 780 square feet. *Ans*

Exercise 27-1   *Every 3*

Express each area as indicated. Round the answers to 2 decimal places.

1. 288 square inches as square feet
2. 86.4 square inches as square feet
3. 45 square feet as square yards
4. 2.25 square feet as square yards
5. 1 600 acres as square miles
6. 192 acres as square miles
7. 130 680 square feet as acres
8. 121 square yards as square rods
9. 17 424 square feet as acres
10. 1 944 square inches as square yards
11. 871 200 square feet as square miles
12. 2 000 square feet as square rods
13. 2.3 square feet as square inches
14. 0.8 square foot as square inches
15. 4 square yards as square feet
16. 0.5 square yard as square feet
17. 3.6 square miles as acres
18. 0.09 square mile as acres
19. 2.1 acres as square feet
20. 0.25 acre as square feet
21. 5 square rods as square yards
22. 0.6 square yard as square inches
23. 0.02 square mile as square feet
24. 1.7 square rods as square feet

Solve each area exercise.

25. How many strips, each having an area of 48 square inches, can be sheared from a sheet of steel which measures 18 square feet?

26. A contractor estimates the cost of developing a 0.3 square mile parcel of land at $1 200 per acre. What is the total cost of developing this parcel?

27. A painter and decorator compute the total interior wall surface of a building as 220 square yards after allowing for windows and doors. Two coats of paint are required for the job. If one gallon of paint covers 500 square feet, how many gallons of paint are required?

28. A bag of lawn food which sells for $16.50 covers 10 000 square feet. What is the cost to the nearer dollar to cover 1 1/2 acres of lawn?

29. A basement floor which measures 900 square feet is to be covered with floor tiles. Each tile measures 100 square inches. Make an allowance of 5% for waste. How many tiles are needed?

METRIC UNITS OF SURFACE MEASURE (AREA)

The method of computing surface measure is the same in the metric system as in the English system. The product of two linear measures produces square measure. The only difference is in the use of metric rather than English units. For example, 2 centimetres X 4 centimetres = 8 square centimetres.

Surface measure symbols are expressed as linear measure symbols with an exponent of 2. For example, 4 square metres is written as 4 m^2, and 25 square centimetres is written as 25 cm^2.

The basic unit of area is the square metre. The surface enclosed by a square which is one metre on a side is one square metre. The surface enclosed by a square which is one centimetre on a side is one square centimetre. Similar meanings are attached to the other square units of measure.

A reduced size drawing of a square decimetre and a square metre is shown. Observe that one linear metre equals 10 linear decimetres, but one square metre equals 10 decimetres X 10 decimetres or 100 square decimetres.

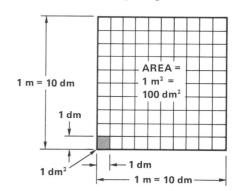

This table shows the units of surface measure with their symbols. These units are based on the square metre. Notice that each unit in the table is 100 times greater than the unit directly above it.

METRIC UNITS OF AREA MEASURE	
1 square millimetre (mm²)	= 0.000 00 1 square metre (m²)
1 square centimetre (cm²)	= 0.000 1 square metre (m²)
1 square decimetre (dm²)	= 0.01 square metre (m²)
1 square metre (m²)	= 1 square metre (m²)
1 square dekametre (dam²)	= 100 square metres (m²)
1 square hectometre (hm²)	= 10 000 square metres (m²)
1 square kilometre (km²)	= 1 000 000 square metres (m²)
1 000 000 square millimetres (mm²)	= 1 square metre (m²)
10 000 square centimetres (cm²)	= 1 square metre (m²)
100 square decimetres (dm²)	= 1 square metre (m²)
1 square metre (m²)	= 1 square metre (m²)
0.01 square dekametre (dam²)	= 1 square metre (m²)
0.000 1 square hectometre (hm²)	= 1 square metre (m²)
0.000 001 square kilometre (km²)	= 1 square metre (m²)

EXPRESSING METRIC AREA MEASURE EQUIVALENTS

To express a given metric unit of area as the next larger metric unit of area, move the decimal point two places to the left. Moving the decimal point two places to the left is actually a short-cut method of dividing by 100.

Examples:

1. Express 840.5 square decimetres (dm²) as square metres (m²).

   Since a square metre is the next larger unit to a square decimetre, move the decimal point 2 places to the left: 8̬40.5

   $$840.5 \text{ dm}^2 = 8.405 \text{ m}^2 \text{ Ans}$$

   In moving the decimal point 2 places to the left, you are actually dividing by 100.

2. Express 3 820 square centimetres (cm²) as square metres (m²).

   Since a square metre is two units larger than a square centimetre, the decimal point is moved 2 × 2 or 4 places to the left.   3̬ 820.

   $$3\ 820 \text{ cm}^2 = 0.382\ 0 \text{ m}^2 \text{ Ans}$$

   In moving the decimal point 4 places to the left, you are actually dividing by 100 × 100 or 10 000.

Section 5 Measure

To express a given metric unit of area as the next smaller metric unit of area, move the decimal point two places to the right. Moving the decimal point two places to the right is actually a short-cut method of multiplying by 100.

Examples:

1. Express 46 square centimetres (cm^2) as square millimetres (mm^2).

    Since a square millimetre is the next smaller unit to a square centimetre, move the decimal point 2 places to the right. 46.00↲

    $$46 \; cm^2 = 4\;600 \; mm^2 \; Ans$$

    In moving the decimal point 2 places to the right, you are actually multiplying by 100.

2. Express 0.08 square kilometre (km^2) as square metres (m^2).

    Since a square metre is three units smaller than a square kilometre, the decimal point is moved 3 × 2 or 6 places to the right.

    0.080 000↲

    $$0.08 \; km^2 = 80\;000 \; m^2 \; Ans$$

    In moving the decimal point 6 places to the right, you are actually multiplying by 100 × 100 × 100 or 1 000 000.

Exercise 27-2 *Every 3*

Express each area as indicated. Round the answers to 3 decimal places when necessary.

1. 500 square millimetres as square centimetres
2. 76 square decimetres as square metres
3. 4 900 square centimetres as square decimetres
4. 15.6 square hectometres as square kilometres
5. 10 000 square millimetres as square decimetres
6. 7 300 square centimetres as square metres
7. 350 000 square millimetres as square metres
8. 9 300 square decimetres as square dekametres
9. 2 700 000 square metres as square kilometres
10. 6.20 square centimetres as square decimetres
11. 800 square decimetres as square hectometres
12. 68 000 square decimetres as square kilometres

13. 8 square metres as square decimetres
14. 23 square centimetres as square millimetres
15. 6.4 square kilometres as square hectometres
16. 134 square decimetres as square centimetres
17. 0.77 square metre as square centimetres
18. 3.4 square decimetres as square millimetres
19. 0.13 square dekametres as square decimetres
20. 0.06 square metre as square millimetres
21. 2.08 square decimetres as square centimetres
22. 0.009 square kilometre as square metres
23. 0.044 square kilometre as square decimetres
24. 0.007 square hectometre as square decimetres

ARITHMETIC OPERATIONS WITH METRIC AREA UNITS

Arithmetic operations are performed with metric area denominate numbers the same as with English area denominate numbers. Compute the arithmetic operations, then write the proper metric unit of surface measure.

Examples:
1. $42 \text{ cm}^2 + 16.5 \text{ cm}^2 = 58.5 \text{ cm}^2$ *Ans*
2. $7.6 \text{ m}^2 - 4.2 \text{ m}^2 = 3.4 \text{ m}^2$ *Ans*
3. $6 \times 30.8 \text{ mm}^2 = 184.8 \text{ mm}^2$ *Ans*
4. $12.6 \text{ km}^2 \div 3 = 4.2 \text{ km}^2$ *Ans*

As with the English system, unlike metric units must be expressed as like units before performing arithmetic operations.

Example:

Determine the total area in square decimetres of these rectangles. The following procedure is the most efficient method to find the solution.

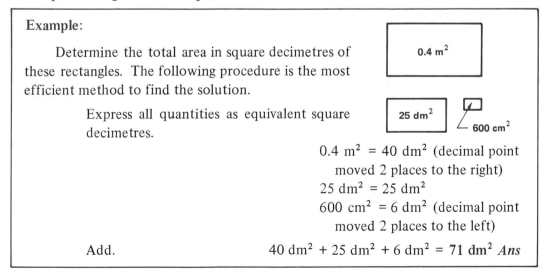

Express all quantities as equivalent square decimetres.

$0.4 \text{ m}^2 = 40 \text{ dm}^2$ (decimal point moved 2 places to the right)
$25 \text{ dm}^2 = 25 \text{ dm}^2$
$600 \text{ cm}^2 = 6 \text{ dm}^2$ (decimal point moved 2 places to the left)

Add. $40 \text{ dm}^2 + 25 \text{ dm}^2 + 6 \text{ dm}^2 = 71 \text{ dm}^2$ *Ans*

Exercise 27-3

Solve each area exercise.

1. How many pieces, each having an area of 280 square centimetres, can be cut from an aluminum sheet which measures 1.4 square metres?
2. A state purchases three parcels of land which are to be developed into a park. The respective areas of the parcels are 16 000 square metres, 21 000 square metres, and 23 000 square metres. How many square kilometres are purchased for the park?
3. Acid soil is corrected (neutralized) by liming. A soil sample shows that 0.4 metric ton of lime per 1 000 square metres of a certain soil is required to correct an acid condition. How many metric tons of lime are needed to neutralize 0.3 square kilometres of soil?
4. An assembly consists of four metal plates. The respective areas of the plates are 500 square centimetres, 700 square centimetres, 18 square decimetres, and 0.15 square metre. Find the total surface measure of the four plates in square metres.
5. A roll of gasket material has a surface measure of 2.25 square metres. Gaskets, each requiring 1 200 square centimetres of material, are cut from the roll. Allow 20% for waste. Find the number of gaskets that can be cut from the roll.

METRIC-ENGLISH AREA MEASURE EQUIVALENTS

Since both the English and metric systems of surface measure are used in this country, it is sometimes necessary to express equivalent measures between systems. The common equivalent surface measures are listed in this table.

AREA MEASURE	
METRIC TO ENGLISH UNITS	
1 square millimetre (mm²)	= 0.001 55 square inch (sq in)
1 square centimetre (cm²)	= 0.155 square inch (sq in)
1 square metre (m²)	= 10.764 square feet (sq ft)
1 square metre (m²)	= 1.196 square yards (sq yd)
1 square kilometre (km²)	= 0.386 1 square mile (sq mi)
ENGLISH TO METRIC UNITS	
1 square inch (sq in)	= 645.2 square millimetres (mm²)
1 square inch (sq in)	= 6.452 square centimetres (cm²)
1 square foot (sq ft)	= 0.092 9 square metre (m²)
1 square yard (sq yd)	= 0.836 square metre (m²)
1 square mile (sq mi)	= 2.589 9 square kilometres (km²)

METHOD 1: To express equivalent units from one system to the other, multiply the given measures by the appropriate equivalent factors listed in the table.

Examples:
1. Express 3 square metres as square feet.
   METHOD 1
   Since 1 m² = 10.764 sq ft, 3 m² = 3 × 10.764 sq ft = **32.292 sq ft** *Ans*

   METHOD 2
   $$\frac{\cancel{3}\,\cancel{m^2}}{1}^{3} \times \frac{10.764 \text{ sq ft}}{\cancel{1\,m^2}_{1}} = 32.292 \text{ sq ft } Ans$$

2. Express 15.80 square inches as square centimetres. Round answer to 2 decimal places.
   Since 1 sq in = 6.452 cm², 15.80 sq in = 15.80 × 6.452 cm² = **101.94 cm²** *Ans*

3. Express 1.2 square kilometres as acres. Round answer to 2 decimal places.
   Express 1.2 square kilometres as square miles. Since 1 km² = 0.386 1 sq mi; 1.2 km² = 1.2 × 0.386 1 sq mi = 0.463 32 sq mi

   Referring to the first table in this unit, 640 acres = 1 square mile. Therefore, 0.463 32 sq mi = 0.463 32 × 640 acres = **296.52 acres** *Ans*

The values in a problem are sometimes given in one system, but the answer is required in a unit of another system. To reduce the amount of calculating, do the required arithmetic operations using the units as given. Then express the answer in units of the other system.

Examples: A job requires four pieces of flat stock of the following areas: 7.8 square feet, 5.2 square feet, 14.6 square feet, and 23.9 square feet. Determine, to 2 decimal places, the total number of square metres of stock required.

Add the values in their given units.

$$7.8 \text{ sq ft} + 5.2 \text{ sq ft} + 14.6 \text{ sq ft} + 23.9 \text{ sq ft} = 51.5 \text{ sq ft}$$

Express the answer, 51.5 square feet, as square metres. Since 1 sq ft = 0.092 9 m², 51.5 sq ft = 51.5 × 0.092 9 m² = **4.78 m²** *Ans*

Exercise 27-4

Express each area as indicated. Round the answers to 3 decimal places.

1. 7 square kilometres as square miles
2. 23 square centimetres as square inches
3. 893 square millimetres as square inches
4. 5.2 square miles as square kilometres
5. 30 square feet as square metres
6. 54 square inches as square centimetres
7. 0.5 square metre as square feet
8. 7 square metres as square yards
9. 2.6 square miles as square kilometres
10. 0.08 square inch as square millimetres
11. 14 square yards as square metres
12. 3 square kilometres as acres
13. 500 acres as square kilometres
14. 1 000 square inches as square metres
15. 0.08 square metre as square inches
16. 0.2 square mile as square metres

Solve each area exercise.

17. A contractor's plywood inventory is as follows: 10 sheets, each sheet having an area of 18 square feet; and 15 sheets, each sheet having an area of 32 square feet. How many total square metres of plywood are in inventory? Express the answer to 1 decimal place.

18. Wall-to-wall carpeting is to be installed in three rooms of a house. The floor area of the rooms are 20 square metres, 27 square metres, and 36 square metres. The carpet installer estimates 10 percent material waste. The cost of carpeting, including installation, is $18.00 per square yard. Find the total cost of carpeting the three rooms.

19. A peach orchard with an area of 0.1 square kilometre contains 3 600 trees. What is the average number of trees per acre? Express the answer to the nearer whole number.

20. Before holes are punched, a sheet metal plate has a surface measure of 450 square centimetres. Four holes, each having an area of 3.60 square inches, and 5 holes, each having an area of 4.74 square inches, are punched in the plate. Find, to 2 decimal places, the number of square centimetres of material remaining in the plate.

21. The interior of a building is to be repainted. Each of two walls has an area of 400 square feet. Each of the other two walls has an area of 650 square feet. A total of 500 square feet is deducted for windows and doorways. One litre of paint covers 28 square metres. How many litres are required for this job? Express the answer to 1 decimal place.

UNIT REVIEW
Exercise 27-5
Express each area as indicated. Round the answers to 2 decimal places.

1. 124 sq in = ? sq ft
2. 13.6 sq ft = ? sq yd
3. 1 200 acres = ? sq mi
4. 108 900 sq ft = ? acres
5. 140 sq yd = ? sq rd
6. 2 300 sq in = ? sq yd
7. 760 000 sq ft = ? sq mi
8. 1 700 sq ft = ? sq rd
9. 0.6 sq ft = ? sq in
10. 1.7 sq yd = ? sq ft
11. 0.06 sq mi = ? acres
12. 0.58 acre = ? sq ft
13. 4 sq rd = ? sq yd
14. 0.5 sq yd = ? sq in
15. 0.03 sq mi = ? sq ft
16. 1.1 sq rd = ? sq ft

Express each area as indicated. Round the answers to 3 decimal places.

17. 650 mm² = ? cm²
18. 3 850 cm² = ? dm²
19. 19 dm² = ? m²
20. 38.7 hm² = ? km²
21. 9 000 mm² = ? dm²
22. 13 500 cm² = ? m²
23. 460 000 m² = ? km²
24. 82 000 mm² = ? m²
25. 47 cm² = ? mm²
26. 230 dm² = ? cm²
27. 5 m² = ? dm²
28. 3.6 km² = ? hm²
29. 1.32 m² = ? cm²
30. 0.9 dm² = ? mm²
31. 0.006 km² = ? m²
32. 0.085 m² = ? mm²
33. 17 cm² = ? sq in
34. 3.8 sq mi = ? km²
35. 570 mm² = ? sq in
36. 0.8 m² = ? sq ft
37. 16 sq in = ? cm²
38. 7.5 sq yd = ? m²
39. 2 km² = ? acres
40. 380 acres = ? km²

PRACTICAL APPLICATIONS
Exercise 27-6

1. How many cardboard cartons can be made from 180 square yards of stock if each carton has a surface measure of 36 square feet?
2. A land developer subdivides a 120-acre tract of land into 160 building lots of approximately the same area. Find the average number of square feet of area per lot.

3. Vinyl tiles are to be installed on a kitchen floor. The area of the floor is 200 square feet. Each tile costs $0.62 and has an area of 100 square inches. Allowing 10% for waste, what is the total tile cost for this job?

4. Before machining, a base plate has an area of 360 square centimetres. Three slots, each having an area of 400 square millimetres; and 2 holes, each having an area of 85 square centimetres, are cut in the plate. What is the area of material remaining in the plate after machining?

5. A 0.15 square kilometre tract of land is to be seeded with Kentucky Bluegrass. At a seeding rate of 0.75 kilograms of seed per 100 square metres of land, how many kilograms of seed are required for the complete tract?

6. A chimney with a total surface area of 13.5 square metres is constructed of brick. Mortar joints between bricks make up 20% of the total chimney surface area. Each brick has a surface area of 18 square inches. How many bricks are required for the chimney?

7. The interior walls of a building are covered with plasterboard. After making allowances for windows and doors, a carpenter estimates that 340 square metres of wall area are covered. One sheet of plasterboard has a surface area of 5.3 square feet. Allowing 15% for waste, how many sheets of plasterboard are required for this job?

8. If 620 kilograms of pesticide are applied to 0.2 square kilometre of farmland, how many pounds of pesticide are applied per acre? One kilogram equals 2.204 6 pounds. Express the answer to one decimal place.

UNIT 28
Equivalent Units of Volume Measure

OBJECTIVES

After studying this unit you should be able to

- Express given English volume measures in larger and smaller units.
- Express given metric volume measures in larger and smaller units.
- Express volume measures as equivalent English and metric units.
- Solve applied problems using English and metric units of volume measure.

ENGLISH UNITS OF VOLUME (CUBIC MEASURE)

A solid is measured by determining the number of cubic units contained in it. A solid is three dimensional; it has length, width, and thickness or height. Length, width, and thickness must be expressed in the same unit of measure. Volume is the product of three linear measures and is expressed in cubic units. For example, 2 inches × 3 inches × 5 inches = 30 cubic inches.

The volume of a cube having sides one foot long is one cubic foot. The volume of a cube having sides one inch long is one cubic inch. A similar meaning is attached to the cubic yard.

A reduced size illustration of a cubic foot and a cubic inch is shown. Observe that 1 linear foot equals 12 linear inches, but 1 cubic foot equals 12 inches × 12 inches × 12 inches or 1 728 cubic inches.

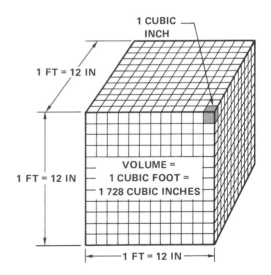

319

This table lists common units of volume measure with their abbreviations. Volume measure units are the same as linear unit measures with the addition of the term "cubic."

ENGLISH UNITS OF VOLUME MEASURE
1 cubic foot (cu ft) = 1 728 cubic inches (cu in)
1 cubic yard (cu yd) = 27 cubic feet (cu ft)

EXPRESSING ENGLISH VOLUME MEASURE EQUIVALENTS

METHOD 1: To express a given unit of volume as a larger unit, divide the given volume by the number of cubic units contained in one of the larger units.

Example: Express 4 320 cubic inches as cubic feet.

METHOD 1
Since 1 728 cu in = 1 cu ft, divide 4 320 by 1 728.
4 320 ÷ 1 728 = 2.5; 4 320 cubic inches = **2.5 cubic feet** *Ans*

METHOD 2
$$\frac{\overset{5}{\cancel{4\,320}\,\cancel{\text{cu in}}}}{1} \times \frac{1 \text{ cu ft}}{\underset{2}{\cancel{1\,728}\,\cancel{\text{cu in}}}} = \mathbf{2.5 \text{ cu ft}} \textit{ Ans}$$

Practical Application Example:

A steel plate has a volume of 1.18 cubic yards. Compute the volume of the plate after 12 000 cubic inches are cut from the plate.

Since 1 728 cubic inches equal 1 cubic foot and 27 cubic feet equal 1 cubic yard, there are 1 728 × 27 or 46 656 cubic inches in 1 cubic yard.

1 728 × 27 = 46 656

Divide 12 000 by 46 656.   12 000 ÷ 46 656 = 0.26

Therefore, 12 000 cubic inches = 0.26 cubic yard.

The volume of the plate after cutting = 1.18 − 0.26 = **0.92 cubic yard.** *Ans*

Unit 28 Equivalent Units of Volume Measure 321

METHOD 1: To express a given unit of volume as a smaller unit, multiply the given volume by the number of cubic units contained in one of the larger units.

Example: Express 9.6 cubic yards as cubic feet.

METHOD 1
Since 27 cu ft = 1 cu yd, multiply 9.6 by 27.
9.6 × 27 = 259.2; 9.6 cubic yards = 259.2 cubic feet *Ans*

METHOD 2
$$\frac{9.6 \cancel{\text{cu yd}}}{1} \times \frac{27 \text{ cu ft}}{1 \cancel{\text{cu yd}}} = 259.2 \text{ cu ft } Ans$$

Practical Application Example:

Castings are to be made from 2 cubic feet of molten metal. If each casting requires 15 cubic inches of metal, how many castings can be made?

Since 1 728 cubic inches equal 1 cubic foot, multiply 1 728 by 2.

1 728 × 2 = 3 456

Therefore, 2 cubic feet = 3 456 cubic inches.

Divide. 3 456 ÷ 15 = 230.4

230 castings can be made. *Ans*

Exercise 28-1

Express each volume as indicated. Round the answers to 2 decimal places.

1. 4 320 cu in = ? cu ft
2. 1 382.4 cu in = ? cu ft
3. 108 cu ft = ? cu yd
4. 8.1 cu ft = ? cu yd
5. 12 900 cu in = ? cu ft
6. 18 600 cu in = ? cu yd
7. 72 cu ft = ? cu yd
8. 124.7 cu ft = ? cu yd
9. 562 cu in = ? cu ft
10. 51 000 cu in = ? cu yd

11. 1.6 cu ft = ? cu in
12. 0.3 cu ft = ? cu in
13. 162 cu yd = ? cu ft
14. 24.3 cu yd = ? cu ft
15. 0.17 cu ft = ? cu in
16. 0.09 cu yd = ? cu in
17. 113.4 cu yd = ? cu ft
18. 10.8 cu yd = ? cu ft
19. 0.55 cu ft = ? cu in
20. 0.03 cu yd = ? cu in

Solve each volume exercise. Round the answers to 2 decimal places.

21. Each casting requires 8.5 cubic inches of bronze. How many cubic feet of molten bronze are required to make 2 500 castings?

22. A cord is a unit of measure of cut and stacked fuel wood equal to 128 cubic feet. Wood is burned at the rate of 1/2 cord per week. How many weeks will a stack of wood measuring 21 1/3 cubic yards last?

23. Common brick weighs 112 pounds per cubic foot. How many cubic yards of brick can be carried by a truck whose maximum carrying load is rated as 8 gross tons? One gross ton = 2 240 pounds.

24. Hot air passes through a duct at the rate of 500 cubic inches per second. Find the number of cubic feet of hot air passing through the duct in one minute.

25. For lumber which is one inch thick, one board foot of lumber has a volume of 1/12 cubic foot. Seasoned white pine weighs 31 1/2 pounds per cubic foot. Find the weight of 1 400 board feet of seasoned white pine.

METRIC UNITS OF VOLUME (CUBIC MEASURE)

The method of computing volume measure is the same in the metric system as in the English system. The product of three linear measures produces cubic measure. The only difference is in the use of metric rather than English units. For example, 2 centimetres × 3 centimetres × 5 centimetres = 30 cubic centimetres.

Volume measure symbols are expressed as linear measure symbols with an exponent of 3. For example, 6 cubic metres is written as 6 m³, and 45 cubic decimetres is written as 45 dm³.

The basic unit of volume is the cubic metre. The volume of a cube having sides

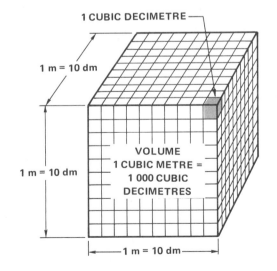

one metre long is one cubic metre. The volume of a cube having sides one decimetre long is one cubic decimetre. Similar meanings are attached to the cubic centimetre and cubic millimetre.

A reduced size illustration of a cubic metre and a cubic decimetre is shown. Observe that 1 linear metre equals 10 linear decimetres, but 1 cubic metre equals 10 decimetres × 10 decimetres × 10 decimetres or 1 000 cubic decimetres.

This table shows the units of volume measure with their symbols. These units are based on the cubic metre. Notice that each unit in the table is 1 000 times greater than the unit directly above it.

METRIC UNITS OF VOLUME MEASURE	
1 cubic millimetre (mm³) = 0.000 000 001 cubic metre (m³)	
1 cubic centimetre (cm³) = 0.000 001 cubic metre (m³)	
1 cubic decimetre (dm³) = 0.001 cubic metre (m³)	
1 cubic metre (m³) = 1 cubic metre (m³)	
1 000 000 000 cubic millimetres (mm³)	= 1 cubic metre (m³)
1 000 000 cubic centimetres (cm³)	= 1 cubic metre (m³)
1 000 cubic decimetres (dm³)	= 1 cubic metre (m³)
1 cubic metre (m³)	= 1 cubic metre (m³)

EXPRESSING METRIC VOLUME MEASURE EQUIVALENTS

To express a given unit of volume as the next larger unit, move the decimal point three places to the left. Moving the decimal point three places to the left is actually a short-cut method of dividing by 1 000.

Examples:

1. Express 1 450 cubic millimetres (mm³) as cubic centimetres (cm³).

    Since a cubic centimetre is the next larger unit to a cubic millimetre, move the decimal point 3 places to the left.   1 450.

    1 450 mm³ = 1.450 cm³ *Ans*

    In moving the decimal point 3 places to the left, you are actually dividing by 1 000.

2. Express 27 000 cubic centimetres (cm³) as cubic metres (m³).

    Since a cubic metre is two units larger than a cubic centimetre, the decimal point is moved 2 × 3 or 6 places to the left.   027 000.

    27 000 cm³ = 0.027 000 m³ *Ans*

    In moving the decimal point 6 places to the left, you are actually dividing by 1 000 × 1 000 or 1 000 000.

To express a given unit of volume as the next smaller unit, move the decimal point three places to the right. Moving the decimal point three places to the right is actually a short-cut method of multiplying by 1 000.

Examples:
1. Express 12.6 cubic metres (m^3) as cubic decimetres (dm^3).

    Since a cubic decimetre is the next smaller unit to a cubic metre, move the decimal point 3 places to the right. 12.600

    $12.6 \ m^3 = 12\ 600 \ dm^3$ *Ans*

    In moving the decimal point 3 places to the right, you are actually multiplying by 1 000.

2. Express 0.08 cubic decimetre (dm^3) as cubic millimetres (mm^3).

    Since a cubic millimetre is two units smaller than a cubic decimetre, the decimal point is moved 2 × 3 or 6 places to the right. 0.080 000

    $0.08 \ dm^3 = 80\ 000 \ mm^3$ *Ans*

    In moving the decimal point 6 places to the right, you are actually multiplying by 1 000 × 1 000 or 1 000 000.

Exercise 28-2

Express each volume as indicated. Round the answers to 2 decimal places.

1. 2 700 mm^3 = ? cm^3
2. 4 320 cm^3 = ? dm^3
3. 940 dm^3 = ? m^3
4. 80 cm^3 = ? dm^3
5. 50 000 mm^3 = ? dm^3
6. 80 cm^3 = ? dm^3
7. 150 000 dm^3 = ? m^3
8. 20 mm^3 = ? cm^3
9. 70 000 mm^3 = ? dm^3
10. 120 000 cm^3 = ? m^3
11. 5 dm^3 = ? cm^3
12. 38 cm^3 = ? mm^3
13. 0.8 m^3 = ? cm^3
14. 0.075 dm^3 = ? cm^3
15. 5.23 cm^3 = ? mm^3
16. 0.94 m^3 = ? dm^3
17. 1.03 dm^3 = ? cm^3
18. 0.089 m^3 = ? cm^3
19. 0.106 dm^3 = ? mm^3
20. 0.006 m^3 = ? cm^3

ARITHMETIC OPERATIONS WITH METRIC VOLUME UNITS

Arithmetic operations are performed with metric volume denominate numbers the same as with English volume denominate numbers. Compute the arithmetic operations, and then write the proper metric unit of volume.

Examples:
1. $4.3 \text{ m}^3 + 11.5 \text{ m}^3 = 15.8 \text{ m}^3$ *Ans*
2. $280 \text{ cm}^3 - 110 \text{ cm}^3 = 170 \text{ cm}^3$ *Ans*
3. $5 \times 1\,400 \text{ mm}^3 = 7\,000 \text{ mm}^3$ *Ans*
4. $12.6 \text{ dm}^3 \div 6 = 2.1 \text{ dm}^3$ *Ans*

As with the English system, unlike units must be expressed as like units before performing arithmetic operations.

Examples: A block has a volume of 0.64 cubic metre. What is the volume of the block in cubic metres after 250 cubic decimetres and 30 000 cubic centimetres of stock are removed?

Express all quantities as cubic metres.

$$250 \text{ dm}^3 = 0.25 \text{ m}^3$$
$$30\,000 \text{ cm}^3 = 0.03 \text{ m}^3$$

Compute the block volume after stock is removed.

$$0.64 \text{ m}^3 - (0.25 \text{ m}^3 + 0.03 \text{ m}^3) =$$
$$0.64 \text{ m}^3 - 0.28 \text{ m}^3 = 0.36 \text{ m}^3 \textit{ Ans}$$

Exercise 28-3

Solve each volume exercise. Round the answers to 2 decimal places.
1. Thirty concrete support bases are required for a construction job. Eighty cubic decimetres of concrete are used for each base. Find the total number of cubic metres of concrete needed for the bases.
2. Before machining, an aluminum piece has a volume of 3.75 cubic decimetres. Machining operations remove 50 cubic centimetres from the top. There are 6 holes drilled. Each hole removes 30 cubic centimetres. There are 4 grooves milled. Each groove removes 2 500 cubic millimetres of stock. Find the volume of the piece, in cubic decimetres, after the machining operations.
3. Anthracite coal weighs 1.5 kilograms per cubic decimetre. Find the weight of a 3.5 cubic metre load of coal.
4. A total of 620 pieces are punched from a strip of stock which has a volume of 38 cubic centimetres. Each piece has a volume of 45 cubic millimetres. How many cubic centimetres of strip stock are wasted after the pieces are punched?
5. A magnesium alloy contains the following volumes of each element: 450 cubic decimetres of magnesium, 42 cubic decimetres of aluminum, 600 cubic centimetres of manganese, and 900 cubic centimetres of zinc. Find the percent composition by volume of each element in the alloy.

METRIC-ENGLISH VOLUME MEASURE EQUIVALENTS

As with linear and surface measure, both the English and metric systems of volume measure are used in this country. Therefore, it is sometimes necessary to express equivalent measures between systems.

The common equivalents of volume measure are listed in the table.

VOLUME MEASURE	
METRIC TO ENGLISH UNITS	
1 cubic centimetre (cm³)	= 0.061 cubic inch (cu in)
1 cubic metre (m³)	= 35.314 cubic feet (cu ft)
1 cubic metre (m³)	= 1.308 cubic yards (cu yd)
ENGLISH TO METRIC UNITS	
1 cubic inch (cu in)	= 16.387 2 cubic centimetres (cm³)
1 cubic foot (cu ft)	= 0.028 3 cubic metre (m³)
1 cubic yard (cu yd)	= 0.764 5 cubic metre (m³)

METHOD 1: To express equivalent units from one system to the other, multiply the given measurements by the appropriate equivalent factors given in the table.

Examples:

1. Express 4.5 cubic metres as cubic yards.

   METHOD 1
   Since 1 m³ = 1.308 cu yd, 4.5 m³ = 4.5 × 1.308 cu yd = 5.886 cu yd *Ans*

   METHOD 2
   $$\frac{4.5 \cancel{m^3}}{1} \times \frac{1.308 \text{ cu yd}}{\cancel{1 m^3}} = 5.886 \text{ cu yd } Ans$$

2. Express 12.63 cubic inches as cubic centimetres. Round the answer to 2 decimal places.

   METHOD 1
   Since 1 cu in = 16.387 2 cm³, 12.63 cu in = 12.63 × 16.387 2 cm³ = 206.97 cm³ *Ans*

   METHOD 2
   $$\frac{12.63 \cancel{\text{cu in}}}{1} \times \frac{16.387\ 2 \text{ cm}^3}{\cancel{1 \text{ cu in}}} = 206.97 \text{ cm}^3 \text{ } Ans$$

The values in a problem are sometimes given in one system, but the answer is required in a unit of another system. To reduce the amount of computing, do the required arithmetic operations using the units as given. Express only the answer in units of the other system.

> Example: The following truckloads of soil are delivered to a construction site: 9 cubic yards, 12 cubic yards, 8 cubic yards, and 10 cubic yards. Determine the total number of cubic metres of soil delivered. Round the answer to 1 decimal place.
>
> Add the values in their given units.
>
> $$9 \text{ cu yd} + 12 \text{ cu yd} + 8 \text{ cu yd} + 10 \text{ cu yd} = 39 \text{ cu yd}$$
>
> Express the answer, 39 cubic yards, as cubic metres.
>
> Since 1 cu yd = 0.764 5 m³, 39 cu yd = 39 × 0.764 5 m³ = 29.8 m³ *Ans*

Exercise 28-4

Express each volume as indicated. Round the answers to 3 decimal places.

1. 9 m³ = ? cu yd
2. 4.6 m³ = ? cu ft
3. 130 cm³ = ? cu in
4. 15 cu yd = ? m³
5. 365 cu ft = ? m³
6. 6.2 cu in = ? cm³
7. 0.3 m³ = ? cu ft
8. 1 260 cu ft = ? m³
9. 1.05 cu yd = ? m³
10. 144 cm³ = ? cu in

Solve each volume exercise. Round the answers to 3 decimal places when necessary.

11. The interior of a building is divided in 5 sections. The sections contain 42 000 cu ft, 37 500 cu ft, 53 000 cu ft, 46 200 cu ft, and 58 600 cu ft. How many total cubic metres of volume are contained in the building?

12. A welded steel spacer plate consists of a bottom plate and 4 spacer blocks. The spacer blocks have volumes of 15.6 cm³, 28.7 cm³, 35 cm³, and 44.6 cm³. The bottom plate volume is 510 cm³.
    a. Find the total volume, in cubic inches, of the welded steel spacer plate.
    b. The steel used for the spacer plate weighs 0.28 pound per cubic inch. Find the total weight of the spacer plate.

13. A building contractor estimates that 750 cubic metres of soil are required to fill and grade a construction site. How many truckloads of soil are hauled to the site if the trucks carry 12 cubic yards per load?

14. If a concrete block contains 440 cubic inches of concrete, how many cubic metres of concrete are required to make 1 500 blocks?

15. Canned goods are packed in a carton which has an inside volume of 0.2 cubic metre. Each can has a volume of 120 cubic inches. Allowing 15% for unused space, how many cans can be packed in the carton?

UNIT REVIEW
Exercise 28-5

Express each volume as indicated. Round the answers to 2 decimal places.

1. 3 800 cu in = ? cu ft
2. 580 cu in = ? cu ft
3. 265 cu ft = ? cu yd
4. 12.6 cu ft = ? cu yd
5. 21 600 cu in = ? cu yd
6. 67 800 cu in = ? cu yd
7. 0.8 cu ft = ? cu in
8. 1.2 cu ft = ? cu in
9. 216 cu yd = ? cu ft
10. 18.9 cu yd = ? cu ft
11. 0.1 cu yd = ? cu in
12. 0.02 cu yd = ? cu in
13. 3 520 mm^3 = ? cm^3
14. 2 910 cm^3 = ? dm^3
15. 870 dm^3 = ? m^3
16. 40 cm^3 = ? dm^3
17. 160 000 mm^3 = ? dm^3
18. 50 000 cm^3 = ? m^3
19. 8 dm^3 = ? cm^3
20. 13 cm^3 = ? mm^3
21. 6.3 m^3 = ? dm^3
22. 24.95 dm^3 = ? cm^3
23. 0.021 m^3 = ? cm^3
24. 0.006 3 dm^3 = ? mm^3

Express each volume as indicated. Round the answers to 3 decimal places.

25. 12 m^3 = ? cu yd
26. 5.8 m^3 = ? cu ft
27. 7.3 cu in = ? cm^3
28. 28 cu yd = ? m^3
29. 205 cm^3 = ? cu in
30. 1 085 cu ft = ? m^3

PRACTICAL APPLICATIONS
Exercise 28-6

Round the answers to 2 decimal places.

1. An aluminum plate with a volume of 0.65 cubic foot is cut into 56 equal pieces. What is the number of cubic inches of volume per piece?

2. Granite weighs 2.7 kilograms per cubic decimetre. Find the number of cubic metres of granite carried by a truck loaded with 6 000 kilograms of granite.

3. A brick wall which will have a volume of 23 cubic metres is to be built. One cubic foot of wall volume requires 20 bricks. Find the total number of bricks needed to construct the wall.

4. A lathe cuts 2.1 cubic inches of stock from a piece in 1 minute. At this speed, how long does it take to cut 0.1 cubic foot of stock from the piece?

5. Two sets of holes are drilled in a brass bracket. In one set, 5 cubic centimetres of stock are removed in drilling each of 3 holes. In the other set, 8 cubic centimetres of stock are removed in drilling each of 6 holes. Before drilling, the bracket has a volume of 700 cubic centimetres of brass. Find the number of cubic inches of brass contained in the bracket after drilling.

6. A manufacturing firm plans to produce 15 000 pieces of a product. Each piece has a volume of 60.5 cubic centimetres of material. Allowing 15% for scrap and waste, how many cubic metres of material are required to produce the 15 000 pieces?

7. For lumber which is 3/4 inch thick, one board foot has a volume of 1/16 cubic foot. Seasoned oak weighs 50 pounds per cubic foot. Compute the weight of 2 300 board feet of 3/4 inch thick seasoned oak.

UNIT 29
Equivalent Units of Capacity and Mass Measure

OBJECTIVE

After studying this unit you should be able to

- Express given English and metric capacity units as larger and smaller units.
- Express capacity measures as equivalent English and metric units.
- Express given English and metric weight units as larger and smaller units.
- Express weight measures as equivalent English and metric units.
- Solve applied problems using English and metric units of capacity and weight measures.

ENGLISH AND METRIC UNITS OF CAPACITY

Capacity is a measure of volume. The capacity of a container is the number of units of material that the container can hold.

In the English system, there are three different kinds of measures of capacity. *Liquid* measure is for measuring liquids. For example, it is used for measuring water and gasoline, and in stating the capacity of fuel tanks and reservoirs. *Dry measure* is for measuring fruit, vegetables, grain, etc.

Apothecaries' fluid measure is for measuring drugs and prescriptions.

The metric system uses only one kind of capacity measure; the units are standardized for all types of measure.

The most common units of English liquid and fluid measure are listed in this table. Also listed are common capacity-cubic measure equivalents.

In the metric system, the litre is the standard unit of capacity. Measures made in gallons in the English system are measured in litres in the metric system. In addition to the litre, the millilitre is used as a unit of capacity measure. Litres and millilitres are used for fluids (gases and liquids) and for dry ingredients in recipes.

ENGLISH UNITS OF CAPACITY MEASURE	
16 ounces (oz)	= 1 pint (pt)
2 pints (pt)	= 1 quart (qt)
4 quarts (qt)	= 1 gallon (gal)
COMMONLY USED CAPACITY–CUBIC MEASURE EQUIVALENTS	
1 gallon (gal)	= 231 cubic inches (cu in)
7.5 gallons (gal)	= 1 cubic foot (cu ft)

The relationship between the litre and millilitre is shown in this table. Also listed are common metric capacity–cubic measure equivalents.

METRIC UNITS OF CAPACITY MEASURE
1 000 millilitres (mL) = 1 litre (L)
COMMONLY USED CAPACITY–CUBIC MEASURE EQUIVALENTS
1 millilitre (mL) = 1 cubic centimetre (cm³)   1 litre (L) = 1 cubic decimetre (dm³)   1 litre (L) = 1 000 cubic centimetres (cm³)   1 000 litres (L) = 1 cubic metre (m³)

EXPRESSING EQUIVALENT ENGLISH CAPACITY MEASURES AND METRIC CAPACITY MEASURES

It is often necessary to express given capacity units as either larger or smaller units. The procedure is the same as used with linear, square, and cubic units of measure.

Examples:

1. Express 20 ounces as pints.

    METHOD 1
    Since 16 oz = 1 pt, divide 20 by 16.
    20 ÷ 16 = 1.25; 20 ounces = **1.25 pints** *Ans*

    METHOD 2
    $$\frac{\overset{5 \text{ oz}}{\cancel{20 \text{ oz}}}}{1} \times \frac{1 \text{ pt}}{\underset{4 \text{ oz}}{\cancel{16 \text{ oz}}}} = 1.25 \text{ pt } Ans$$

2. How many ounces of solution are contained in a 3/4-quart container when full?

    METHOD 1
    Since 16 oz = 1 pt and 2 pt = 1 qt, there are 16 × 2 or 32 ounces in 1 quart. Multiply 32 by 3/4.

    $32 \times \frac{3}{4} = 24$; $\frac{3}{4}$-quart = **24 ounces** *Ans*

    METHOD 2
    $$\frac{\overset{0.75}{\cancel{0.75 \text{ qt}}}}{1} \times \frac{\overset{2}{\cancel{2 \text{ pt}}}}{\underset{1}{\cancel{1 \text{ qt}}}} \times \frac{16 \text{ oz}}{\underset{1}{\cancel{1 \text{ pt}}}} = 24 \text{ oz } Ans$$

Practice

1. Express 2.4 litres as millilitres. 2 400 mL *Ans*
2. Express 300 millilitres as litres. 0.300 L *Ans*
3. A storage tank contains 3 200 litres of fuel oil when one-third full. What is the total volume of the tank in cubic metres? 9.600 m³ *Ans*
4. An automobile gasoline tank is designed to hold 18 gallons of gasoline. How many cubic feet of space does the tank contain? The tank contains 2.4 cu ft *Ans*

Exercise 29-1

Express each unit of measure as indicated. Round the answers to 2 decimal places.

1. 6.5 pt = ? qt
2. 30 oz = ? pt
3. 18 gal = ? qt
4. 0.4 pt = ? oz
5. 35 qt = ? gal
6. 17 qt = ? pt
7. 3.2 gal = ? cu in
8. 23 gal = ? cu ft
9. 50 cu in = ? gal
10. 80 cu ft = ? gal
11. 62 oz = ? qt
12. 0.2 gal = ? pt
13. 1.6 qt = ? oz
14. 43 pt = ? gal
15. 2 700 mL = ? L
16. 1.2 L = ? mL
17. 23.6 mL = ? cm³
18. 3.9 L = ? cm³
19. 5 300 cm³ = ? L
20. 218 cm³ = ? mL
21. 0.08 m³ = ? L
22. 650 L = ? m³
23. 83 dm³ = ? L
24. 0.63 L = ? mL
25. 9.6 dm³ = ? L
26. 478 mL = ? L
27. 29 000 mL = ? L
28. 750 L = ? m³

Solve each exercise

29. In planning for a reception, a chef estimates that eighty 4-ounce servings of tomato juice are required. How many quarts of juice must be ordered?

30. The liquid intake of a hospital patient during a specified period of time is 300 mL, 250 mL, 125 mL, 275 mL, 350 mL, 150 mL, and 200 mL. Find the total litre intake of liquid for this time period.

31. A water tank has a volume of 4 500 cubic feet. The tank is 9/10 full. How many gallons of water are contained in the tank?

32. An automobile engine originally has a displacement of 2.3 litres. The engine is rebored an additional 150 cubic centimetres. What is the engine displacement in litres after it is rebored?

33. An empty fuel oil tank has a volume of 400 cubic feet. Oil is pumped into the tank at the rate of 75 gallons per minute. How long does it take to fill the tank?

34. A bottle contains 2.250 litres of solution. A laboratory technician takes 28 samples from the bottle. Each sample contains 35 millilitres of solution. How many litres of solution remain in the bottle?

35. An engine running at a constant speed uses 120 millilitres of gasoline per minute. How many litres of gasoline are used in 5 hours?

36. A solution contains 10% acid and 90% water. How many quarts of solution are made with 5 ounces of acid? Round the answer to 2 decimal places.

METRIC-ENGLISH CAPACITY UNIT EQUIVALENTS

Because both English and metric capacity measures are used in this country, it is sometimes necessary to express equivalent units from one system to the other.

The most common equivalents of capacity measure are listed in the table.

CAPACITY MEASURE	
Metric to English Units	
1 millilitre (mL)	= 0.034 ounce (oz)
1 litre (L)	= 1.057 quarts (qt)
1 litre (L)	= 0.264 gallon (gal)
English to Metric Units	
1 ounce (oz)	= 29.563 millilitres (mL)
1 quart (qt)	= 0.946 litre (L)
1 gallon (gal)	= 3.785 litres (L)

METHOD 1: To express units from one system as units in the other system, multiply the given measurements by the appropriate equivalent factors listed in the table.

Examples:

1. Express 2.50 ounces as millilitres. Round the answer to 2 decimal places.
   METHOD 1
   Since 1 oz = 29.563 mL, 2.50 oz = 2.50 × 29.563 mL = 73.91 mL *Ans*

   METHOD 2
   $$\frac{2.50 \text{ oz}}{1} \times \frac{29.563 \text{ mL}}{1 \text{ oz}} = 73.91 \text{ mL } Ans$$

2. Express 28.65 litres as gallons. Round the answer to 2 decimal places.
   METHOD 1
   Since 1 L = 0.264 gal, 28.65 L = 28.65 × 0.264 gal = 7.56 gal *Ans*

   METHOD 2
   $$\frac{28.65 \text{ L}}{1} \times \frac{0.264 \text{ gal}}{1 \text{ L}} = 7.56 \text{ gal } Ans$$

Section 5 Measure

The values in capacity problems are sometimes given in one system, but the answer is required in a unit of the other system. To reduce the amount of computing, do the arithmetic operations using the units as given. Express only the answer as units in the other system.

> Example: The following amounts of fuel oil are delivered to customers: 190 gallons, 275 gallons, 210 gallons, and 172 gallons. How many litres of oil are delivered? Round the answer to the nearer litre.
>
> Add the values in their given units.
>
> $$190 \text{ gal} + 275 \text{ gal} + 210 \text{ gal} + 172 \text{ gal} = 847 \text{ gal}$$
>
> Express 847 gallons as litres.
>
> $$\text{Since } 1 \text{ gal} = 3.785 \text{ L}, 847 \times 3.785 \text{ L} = 3\,206 \text{ L } Ans$$

Exercise 29-2

Express each unit of capacity as indicated. Round the answers to 3 decimal places.

1. 16 L = ? gal
2. 5 gal = ? L
3. 2.6 oz = ? mL
4. 8.5 L = ? qt
5. 1.38 qt = ? L
6. 132 mL = ? oz
7. 570 L = ? gal
8. 0.75 qt = ? L
9. 9.8 gal = ? L
10. 0.6 oz = ? mL
11. 0.430 L = ? pt
12. 15 oz = ? L
13. 0.37 L = ? oz
14. 1.6 pt = ? L

Solve each exercise. Round the answer to 2 decimal places.

15. A 2-quart container is 2/3 full. Find the number of litres of lubricating oil in the container.
16. A pump discharges 160 litres of water per minute. The pump operated for 2 1/2 hours. How many gallons of water are discharged?
17. The following amounts of acid are removed from a full 1-pint container: 40 mL, 35 mL, 45 mL, and 50 mL. How many ounces of acid remain in the container?
18. A tank has a volume of 5 cubic metres. How many gallons of oil can be stored in the tank?
19. The average highway mileage for a certain automobile is 34 miles per gallon of diesel fuel. Compute the car's mileage in kilometres per litre of fuel. One kilometre equals 0.621 4 mile.

ENGLISH AND METRIC UNITS OF WEIGHT (MASS)

Weight is a measure of the force of attraction of the earth on an object. *Mass* is a measure of the amount of matter contained in an object. The weight of an object varies with its distance from the earth's center. The mass of an object remains the same regardless of its location in the universe.

Scientific applications dealing with objects located other than on the earth's surface are *not* considered in this book. Therefore, the terms weight and mass are used interchangeably.

As with capacity measures, the English system has three types of weight measures. *Troy weights* are used in weighing jewels and precious metals such as gold and silver. *Apothecaries' weights* are for measuring drugs and prescriptions. *Avoirdupois* or *commercial weights* are used for all commodities except precious metals, jewels, and drugs.

The most common units of English weight measure are listed in this table.

ENGLISH UNITS OF WEIGHT MEASURE	
16 ounces (oz)	= 1 pound (lb)
2 000 pounds (lb)	= 1 net or short ton
2 240 pounds (lb)	= 1 gross or long ton

In the metric system, the kilogram is the standard unit of mass. Objects that are measured in pounds in the English system are measured in kilograms in the metric system.

The most common units of metric weight (mass) are listed in this table.

METRIC UNITS OF WEIGHT (MASS) MEASURE	
1 000 milligrams (mg)	= 1 gram (g)
1 000 grams (g)	= 1 kilogram (kg)
1 000 kilograms (kg)	= 1 metric ton (t)

EXPRESSING EQUIVALENT ENGLISH WEIGHT MEASURES AND METRIC WEIGHT MEASURES

In both the English and metric systems, apply the same procedures that are used with other measures. The following examples show the method of expressing given weight units as larger or smaller units.

Examples:

1. What is the total weight in pounds of 80 electrical switches if each switch weighs 2.5 ounces?

    Compute the total weight in ounces. $80 \times 2.5 \text{ oz} = 200 \text{ oz}$

    Express 200 ounces as pounds.

    METHOD 1
    Since 16 oz = 1 lb, divide 200 by 16.
    $200 \div 16 = 12.5$; 200 ounces = **12.5 lb** *Ans*

    METHOD 2
    $$\frac{\cancel{200}^{25} \text{ oz}}{1} \times \frac{1 \text{ lb}}{\cancel{16}_{2} \text{ oz}} = 12.5 \text{ lb } Ans$$

2. A shipment of 500 forgings weighs 1.6 metric tons. What is the weight of each forging in kilograms?

    Compute the weight per forging in metric tons.
    $$1.6 \text{ t} \div 500 = 0.003\,2 \text{ t}$$

    Since 1 metric ton equals 1 000 kilograms, move the decimal point 3 places to the right. $0.\underline{003\,2}$

    The weight of each forging = **3.2 kg** *Ans*

Exercise 29-3

Express each unit of weight as indicated. Round the answers to 2 decimal places.

1. 35 oz = ? lb
2. 0.6 lb = ? oz
3. 2.4 long tons = ? lb
4. 3.1 short tons = ? lb
5. 5 300 lb = ? short tons
6. 7 850 lb = ? long tons
7. 43.5 oz = ? lb
8. 0.05 lb = ? oz
9. 0.12 short tons = ? lb
10. 720 lb = ? long tons
11. 1.72 g = ? mg
12. 890 mg = ? g
13. 2.6 metric tons = ? kg
14. 1 230 g = ? kg
15. 2 700 kg = ? metric tons
16. 0.6 kg = ? g
17. 0.04 g = ? mg
18. 900 kg = ? metric tons
19. 23 000 mg = ? g
20. 80 g = ? kg

Solve each exercise. Round the answers to 2 decimal places.

21. What is the total weight, in pounds, of 1 gross (144) 12-ounce cans of fruit?

22. An analytical balance is used by laboratory technicians in measuring the following weights: 750 mg, 600 mg, 920 mg, 550 mg, and 870 mg. Find the total measured weight in grams.

23. Three hundred identical strips are sheared from a sheet of steel. The sheet weighs 18.3 kilograms. Find the weight, in grams, of each strip.

24. Aluminum weighs 2.707 metric tons per cubic metre. How many kilograms does 0.05 cubic metre of aluminum weigh?

25. A force of 760 tons (short tons) is exerted on the base of a steel support column. The base has a cross-sectional area of 160 square inches. How many pounds of force are exerted per square inch of cross-sectional area?

26. A truck delivers 6 prefabricated concrete wall sections to a job site. Each wall section has a volume of 0.75 cubic metre. One cubic metre of concrete weighs 2 300 kilograms. Find the number of metric tons in this delivery.

27. An assembly housing weighs 8.25 pounds. The weight of the housing is reduced to 6.50 pounds by drilling holes in the housing. Each drilled hole removes 0.80 ounce of material. How many holes are drilled?

28. Water weighs 62.42 pounds per cubic foot at 40 degrees Fahrenheit and 61.74 pound per cubic foot at 120 degrees Fahrenheit. Over this temperature change, what is the average decrease in weight in ounces per cubic foot for each degree increase in temperature?

METRIC-ENGLISH WEIGHT (MASS) UNIT EQUIVALENTS

As with other measures, the United States uses a dual system of weight measure. It is sometimes necessary to express units from one system as units in the other system. The most common equivalents of weight measure are listed in this table.

WEIGHT MEASURE	
Metric to English Units	English to Metric Units
1 gram (g) = 0.035 ounce (oz) 1 kilogram (kg) = 2.205 pounds (lb) 1 metric ton (t) = 1.102 short tons 1 metric ton (t) = 0.984 long ton	1 ounce (oz) = 28.348 grams (g) 1 pound (lb) = 0.454 kilogram (kg) 1 short ton = 0.907 metric ton (t) 1 long ton = 1.016 metric tons (t)

338 Section 5 Measure

METHOD 1: To express equivalent units from one system to the other, multiply the given measures by the appropriate equivalent factors listed in the table.

Examples:

1. Express 35 pounds as kilograms.

   METHOD 1
   Since 1 lb = 0.454 kg, 35 lb = 35 × 0.454 kg = 15.89 kg *Ans*

   METHOD 2
   $$\frac{\cancel{35 \text{ lb}}^{35}}{1} \times \frac{0.454 \text{ kg}}{\cancel{1 \text{ lb}}_{1}} = 15.89 \text{ kg } Ans$$

2. Express 8.26 metric tons as short tons. Round the answer to 2 decimal places.

   METHOD 1
   Since 1 metric ton = 1.102 short tons, 8.26 metric tons = 8.26 × 1.102 short tons = **9.10 short tons** *Ans*

   METHOD 2
   $$\frac{\cancel{8.26 \text{ metric tons}}^{8.26}}{1} \times \frac{1.102 \text{ short tons}}{\cancel{1 \text{ metric ton}}_{1}} = 9.10 \text{ short tons } Ans$$

The values in weight problems are sometimes given in one system, but the answer is required in a unit of the other system. To reduce the amount of computing, do the required arithmetic operations using the units as given. Express only the answer as units in the other system.

Example: Determine the total weight in ounces of 7 grams, 10.5 grams, 22 grams, and 18 grams. Round the answer to 2 decimal places.

Add the values in their given units.   7 g + 10.5 g + 22 g + 18 g = 57.5 g

Express 57.5 grams as ounces.

Since 1 g = 0.035 oz, 57.5 g = 57.5 × 0.035 oz = **2.01 oz** *Ans*

Exercise 29-4

Express each unit of measure as indicated. Round the answers to 3 decimal places.

1. 37 g = ? oz
2. 0.8 oz = ? g
3. 16.6 metric tons = ? short tons
4. 103 kg = ? lb
5. 80 long tons = ? metric tons
6. 793.6 lb = ? kg
7. 0.95 kg = ? lb
8. 57.2 short tons = ? metric tons
9. 322 g = ? oz
10. 20 oz = ? kg
11. 8 720 lb = ? metric tons
12. 1.74 metric tons = ? lb

Solve each exercise. Round the answers to 2 decimal places.

13. A hospital dietitian specifies 60 g, 115 g, and 130 g of meat daily for a patient. How many total ounces of meat are allowed for the day?

14. Before machining operations, a forging weighed 3.75 pounds. Three operations remove the following amounts of material: 10 oz, 8.5 oz, and 11.2 oz. What is the weight of the forging, in kilograms, after the machining operations?

15. A roll of 0.042 inch thick sheet copper has an area of 18.75 square feet. The copper sheet weighs 2.5 pounds per square foot. Find the weight, in kilograms, of the roll.

16. Granite weighs 2 693.5 kilograms per cubic metre. Find the weight of granite in pounds per cubic foot.

UNIT REVIEW

Exercise 29-5

Express each unit of measure as indicated. Round the answers to 2 decimal places.

1. 25 oz = ? pt
2. 13 gal = ? qt
3. 7.25 pt = ? qt
4. 28 qt = ? gal
5. 2.7 gal = ? cu in
6. 92 cu ft = ? gal
7. 0.4 gal = ? pt
8. 1.2 qt = ? oz
9. 2.6 L = ? mL
10. 3 850 mL = ? L
11. 87.9 mL = ? cm^3
12. 4 420 cm^3 = ? L
13. 526 L = ? m^3
14. 4.32 dm^3 = ? L
15. 0.09 L = ? mL
16. 26 000 mL = ? L

Express each unit of measure as indicated. Round the answers to 3 decimal places.

17. 7 gal = ? L
18. 9.6 L = ? qt
19. 3.4 oz = ? mL
20. 2.63 qt = ? L

21. 97 mL = ? oz
22. 342 L = ? gal
23. 0.58 L = ? pt
24. 19 oz = ? L

Express each unit of weight as indicated. Round the answers to 2 decimal places.
25. 0.8 lb = ? oz
26. 43 oz = ? lb
27. 6 700 lb = ? short tons
28. 1.8 long tons = ? lb
29. 93 000 lb = ? long tons
30. 0.73 short tons = ? lb
31. 930 mg = ? g
32. 2 730 g = ? kg
33. 3.4 metric tons = ? kg
34. 0.85 kg = ? g
35. 3.06 g = ? mg
36. 230 000 kg = ? metric tons

Express each unit of weight as indicated. Round the answer to 3 decimal places.
37. 1.3 oz = ? g
38. 235 kg = ? lb
39. 430 g = ? oz
40. 105.3 lb = ? kg
41. 63.9 short tons = ? metric tons
42. 2.84 metric tons = ? lb

PRACTICAL APPLICATIONS

Exercise 29-6

Round the answers to 2 decimal places.

1. A laboratory technician withdraws the following amounts of solution from a full 2-litre container: 270 mL, 95 mL, 370 mL, and 140 mL. How many litres of solution remain in the container?

2. A water storage tank has a volume of 10 800 cubic feet. The tank is 4/5 full. How many gallons of water does the tank contain?

3. An automobile weighs 1.680 metric tons. Find the average weight, in kilograms, exerted on each of its wheels.

4. A baker prepares an order to make 60 dozen rolls. Each roll weighs 3.75 ounces. How many pounds of dough must the baker prepare?

5. Air-dried, commercial white fir lumber weighs 27 pounds per cubic foot. What is the weight, in long tons, of a shipment of 450 cubic feet of white fir?

6. An engine uses 24 litres of gasoline while running for 3.2 hours. Find the average gas consumption for the engine in millilitres per minute.

7. Each electronic pocket calculator weighs 336 grams. The calculators are packed in a carton which weighs 2.25 kilograms. Find the weight, in kilograms, if each carton contains 200 calculators.

8. How many quarts of water must be added to 3 ounces of acid to make a solution which is 4% acid and 96% water?

Appendix

DECIMAL EQUIVALENT TABLE

1/64 — 0.015 625	33/64 – 0.515 625
1/32 — 0.031 25	17/32 — 0.531 25
3/64 — 0.046 875	35/64 – 0.546 875
1/16 — 0.062 5	9/16 — 0.562 5
5/64 — 0.078 125	37/64 – 0.578 125
3/32 — 0.093 75	19/32 — 0.593 75
7/64 — 0.109 375	39/64 – 0.609 375
1/8 — 0.125	5/8 — 0.625
9/64 — 0.140 625	41/64 – 0.640 625
5/32 — 0.156 25	21/32 — 0.656 25
11/64 – 0.171 875	43/64 – 0.671 875
3/16 — 0.187 5	11/16 — 0.687 5
13/64 – 0.203 125	45/64 – 0.703 125
7/32 — 0.218 75	23/32 — 0.718 75
15/64 – 0.234 375	47/64 – 0.734 375
1/4 — 0.25	3/4 — 0.75
17/64 – 0.265 625	49/64 – 0.765 625
9/32 — 0.281 25	25/32 — 0.781 25
19/64 – 0.296 875	51/64 – 0.796 875
5/16 — 0.312 5	13/16 — 0.812 5
21/64 – 0.328 125	53/64 – 0.828 125
11/32 — 0.343 75	27/32 — 0.843 75
23/64 – 0.359 375	55/64 – 0.859 375
3/8 — 0.375	7/8 — 0.875
25/64 – 0.390 625	57/64 – 0.890 625
13/32 — 0.406 25	29/32 — 0.906 25
27/64 – 0.421 875	59/64 – 0.921 875
7/16 — 0.437 5	15/16 — 0.937 5
29/64 – 0.453 125	61/64 – 0.953 125
15/32 — 0.468 75	31/32 — 0.968 75
31/64 – 0.484 375	63/64 – 0.984 375
1/2 — 0.5	1 — 1.

LINEAR MEASURE

ENGLISH UNITS OF LINEAR MEASURE

1 foot (ft)	=	12 inches (in)
1 yard (yd)	=	3 feet (ft)
1 yard (yd)	=	36 inches (in)
1 rod (rd)	=	16.5 feet (ft)
1 furlong	=	220 yards (yd)
1 mile (mi)	=	5 280 feet (ft)
1 mile (mi)	=	1 760 yards (yd)
1 mile (mi)	=	320 rods (rd)
1 mile (mi)	=	8 furlongs

METRIC UNITS OF LINEAR MEASURE

10 millimetres (mm)	=	1 centimetre (cm)
10 centimetres (cm)	=	1 decimetre (dm)
10 decimetres (dm)	=	1 metre (m)
10 metres (m)	=	1 dekametre (dam)
10 dekametres (dam)	=	1 hectometre (hm)
10 hectometres (hm)	=	1 kilometre (km)

METRIC TO ENGLISH UNITS

1 millimetre (mm)	=	0.039 37 inch (in)
1 centimetre (cm)	=	0.393 7 inch (in)
1 metre (m)	=	39.37 inches (in)
1 metre (m)	=	3.280 8 feet (ft)
1 kilometre (km)	=	0.621 4 mile (mi)

ENGLISH TO METRIC UNITS

1 inch (in)	=	25.4 millimetres (mm)
1 inch (in)	=	2.54 centimetres (cm)
1 foot (ft)	=	0.304 8 metre (m)
1 yard (yd)	=	0.914 4 metre (m)
1 mile (mi)	=	1.609 kilometres (km)

AREA MEASURE

ENGLISH UNITS OF AREA MEASURE

1 square foot (sq ft)	=	144 square inches (sq in)
1 square yard (sq yd)	=	9 square feet (sq ft)
1 square rod (sq rd)	=	30.25 square yards (sq yd)
1 acre (A)	=	160 square rods (sq rd)
1 acre (A)	=	43 560 square feet (sq ft)
1 square mile (sq mi)	=	640 acres (A)

METRIC UNITS OF AREA MEASURE

100 square millimetres (mm^2)	=	1 square centimetre (cm^2)
100 square centimetres (cm^2)	=	1 square decimetre (dm^2)
100 square decimetres (dm^2)	=	1 square metre (m^2)
100 square metres (m^2)	=	1 square dekametre (dam^2)
100 square dekametres (dam^2)	=	1 square hectometre (hm^2)
100 square hectometres (hm^2)	=	1 square kilometre (km^2)

METRIC TO ENGLISH UNITS

1 square millimetre (mm^2)	=	0.001 55 square inch (sq in)
1 square centimetre (cm^2)	=	0.155 square inch (sq in)
1 square metre (m^2)	=	10.764 square feet (sq ft)
1 square metre (m^2)	=	1.196 square yards (sq yd)
1 square kilometre (km^2)	=	0.386 1 square mile (sq mi)

ENGLISH TO METRIC UNITS

1 square inch (sq in)	=	645.2 square millimetres (mm^2)
1 square inch (sq in)	=	6.452 square centimetres (cm^2)
1 square foot (sq ft)	=	0.092 9 square metre (m^2)
1 square yard (sq yd)	=	0.836 square metre (m^2)
1 square mile (sq mi)	=	2.589 9 square kilometres (km^2)

VOLUME MEASURE

ENGLISH UNITS OF VOLUME MEASURE

1 cubic foot (cu ft) = 1 728 cubic inches (cu in)
1 cubic yard (cu yd) = 27 cubic feet (cu ft)

METRIC UNITS OF VOLUME MEASURE

1 000 cubic millimetres (mm^3) = 1 cubic centimetre (cm^3)
1 000 cubic centimetres (cm^3) = 1 cubic decimetre (dm^3)
1 000 cubic decimetres (dm^3) = 1 cubic metre (m^3)
1 000 cubic metres (m^3) = 1 cubic dekametre (dam^3)
1 000 cubic dekametres (dam^3) = 1 cubic hectometre (hm^3)
1 000 cubic hectometres (hm^3) = 1 cubic kilometre (km^3)

METRIC TO ENGLISH UNITS

1 cubic centimetre (cm^3) = 0.061 cubic inch (cu in)
1 cubic metre (m^3) = 35.314 cubic feet (cu ft)
1 cubic metre (m^3) = 1.308 cubic yards (cu yd)

ENGLISH TO METRIC UNITS

1 cubic inch (cu in) = 16.387 2 cubic centimetres (cm^3)
1 cubic foot (cu ft) = 0.028 3 cubic metre (m^3)
1 cubic yard (cu yd) = 0.764 5 cubic metre (m^3)

CAPACITY MEASURE

ENGLISH UNITS OF CAPACITY MEASURE

16 ounces (oz) = 1 pint (pt)
2 pints (pt) = 1 quart (qt)
4 quarts (qt) = 1 gallon (gal)

COMMONLY USED CAPACITY -- CUBIC MEASURE EQUIVALENTS

1 gallon (gal) = 231 cubic inches (cu in)
7.5 gallons (gal) = 1 cubic foot (cu ft)

METRIC UNITS OF CAPACITY MEASURE

10 millilitres (mL) = 1 centilitre (cL)
10 centilitres (cL) = 1 decilitre (dL)
10 decilitres (dL) = 1 litre (L)
10 litres (L) = 1 dekalitre (daL)
10 dekalitres (daL) = 1 hectolitre (hL)
10 hectolitres (hL) = 1 kilolitre (kL)

COMMONLY USED CAPACITY -- CUBIC MEASURE EQUIVALENTS

1 millilitre (mL) = 1 cubic centimetre (cm^3)
1 litre (L) = 1 cubic decimetre (dm^3)
1 litre (L) = 1 000 cubic centimetres (cm^3)
1 000 litres (L) = 1 cubic metre (m^3)

METRIC TO ENGLISH UNITS

1 millilitre (mL) = 0.034 ounce (oz)
1 litre (L) = 1.057 quarts (qt)
1 litre (L) = 0.264 gallon (gal)

ENGLISH TO METRIC UNITS

1 ounce (oz) = 29.563 millilitres (mL)
1 quart (qt) = 0.946 litre (L)
1 gallon (gal) = 3.785 litres (L)

WEIGHT (MASS) MEASURE

ENGLISH UNITS OF WEIGHT MEASURE

16 ounces (oz)	=	1 pound (lb)
2 000 pounds (lb)	=	1 net or short ton
2 240 pounds (lb)	=	1 gross or long ton

METRIC UNITS OF WEIGHT (MASS) MEASURE

10 milligrams (mg)	=	1 centigram (cg)
10 centigrams (cg)	=	1 decigram (dg)
10 decigrams (dg)	=	1 gram (g)
10 grams (g)	=	1 dekagram (dag)
10 dekagrams (dag)	=	1 hectogram (hg)
10 hectograms (hg)	=	1 kilogram (kg)
1 000 kilograms (kg)	=	1 metric ton (t)

METRIC TO ENGLISH UNITS

1 gram (g)	=	0.035 ounce (oz)
1 kilogram (kg)	=	2.205 pounds (lb)
1 metric ton (t)	=	1.102 short tons
1 metric ton (t)	=	0.984 long ton

ENGLISH TO METRIC UNITS

1 ounce (oz)	=	28.348 grams (g)
1 pound (lb)	=	0.454 kilogram (kg)
1 short ton	=	0.907 metric ton (t)
1 long ton	=	1.016 metric tons (t)

Index

A

Addends, 8
Addition, common fractions, 66-78, 83, 96, 108
    compound numbers, 280-281
    decimal fractions, 141-143, 145, 157, 170
    whole numbers, 8-13
Apothecaries', fluid measure, 330
Apothecaries' weights, 335
Area measure, 306-317
    English units, 306-310
    metric-English equivalents, 314-317
    metric units, 310-314
    powers and roots, 176
    tables, 343
Arithmetic operations, English linear measure, 280-285
    metric area units, 313-314
    metric volume units, 324-325
Arithmetic operations with compound numbers, 280-285, 295-296
Avoirdupois weights, 335

B

Bar graphs, 240-250
    drawing, 246-250
    reading, 241-246
Base, percentage, 212
    finding, 216-218
    finding percentage by, 213, 215
Basic operations, 1
Broken-line graph, 253
    drawing, 262-264

C

Capacity measure, tables, 345
    units of, 330-334
Causal relationships, graphs, 253-254
Central tendency measures, 228
    mean, 228-230
    median, 231-233
    mode, 233-235
Circle, formula, 46
Combined data line graphs, reading, 256-261
Combined operations, decimal fractions, 145, 157, 170
    fractions, 116-124
    powers, 179
    roots, 186

Commercial weights, 335
Common fractions, 51-65
    addition, 66-78, 96
    as decimal fractions, 134-137
    as mixed numbers, 59
    as percents, 210
    combined operations, 116-124
    comparing values, 69
    decimal fractions as, 137-138
    division, 103-115
    equivalents, 54
    lowest terms, 56
    mixed numbers, 58
    multiplication, 90-102
    parts, 52
    percents as, 211
    powers, 179
    subtraction, 79-89, 96
Compound numbers, arithmetic operations with, 280-285
Coordinate paper, 240
Counting numbers, 4
Cross-section graph paper, 240
Cube root, 180
Cubic measure, English, 319-322
    metric, 322-324
Curved-line graphs, 254
    drawing, 265-268
Cylinder wall, 304

D

Data, statistical measures, 228-235
Decimal equivalent table, 193-194, 341
Decimal fractions, 125-129
    addition and subtraction, 141-145
    as common fractions, 137-138
    as percents, 209-210
    combined operations, 193-201
    common fractions as, 134-137
    division, 165-170
    equivalents, 132-138
    multiplication, 152-159
    percents as, 211
    powers and roots, 176-187
    reading, 127-129
    writing, 129
Decimal system, 1
    place value, 2
Denominate numbers, 274
Denominator, 34
    fractions, 52

Difference, subtraction, 14
Digits, 2
    space between groups of three, 294
Dispersion of data, 228
    mean deviation, 236-237
    range, 235-236
Dividend, 34
Division, compound numbers, 284-285
    decimal fractions, 165-170
    fractions, 56, 103-115
        combined operations, 116-123
    indicated by fraction, 54
    whole numbers, 33-42
Divisor, 34
Drawing bar graph, 246
Drawing line graph, 262-268
Dry measure, defined, 330

E

Electrical field, formula, 43
    formula for current, 272
English linear measure, 274-285
    metric equivalents, 296-299
English units, of area measurement, 306-317
    of capacity measure, 330
        metric equivalents, 333
    of surface measure, 306-316
        metric equivalents, 314-317
    of volume measure, 319-322
        metric equivalents, 326
    of weight measure, 335
        metric equivalents, 326, 337
Equivalent fractions, 54
    decimal and common, 132-138
    decimal equivalent table, 194
Equivalent units of measure, 275
    capacity and mass measure, 330-339
    metric system, 294
    surface measure, 314
    tables, 342-346
    volume measure, 320, 323, 326
Expanding whole numbers, 3
Exponent, 176

F

Factors, defined, 66, 176
    multiplying by three or more, 26, 154
Feeler gauge, metric, 303
Fluid measure, apothecaries', 330
    See also Volume

347

Formulas, 43
   area, 124
   circle, 46
   electric current, 182, 272
      resistance, 121
   electronics, 50
   horsepower, 49, 121
   power, 43
   temperature, 46, 48
   weight, 124
   work, 272
Fractional dimensions, 193
Fractional parts, 52
   meaning, 126
Fractions, defined, 51
   See also Common fractions; Decimal fractions
Frequency distribution table, 233-234

G
Graphs, defined, 240
   bar, 240-250
   line, 253-268
   types and structure, 240
Grouping symbols, 43, 178, 182

H
Horizontal scale, graphs, 240

I
Inch scale, 152, 166
Index, roots, 180

L
Lengths, measurement, 274
   See also Linear measure
Linear measure, English units, 274-285
   metric units, 291-297
   tables, 342
Line graphs, 253-268
   drawing, 262-268
      broken-line graphs, 262-264
      curved-line graphs, 265-268
      straight-line graphs, 264-265
      reading, 254-261
Liquid measure, defined, 330
   units of capacity measure, 330-335
Litres, 330-331
Lockplate, 304
Long division symbol, 34
Long multiplication, 24
Lowest common denominator, 66-68

M
Mass, English and metric units of, 335-339
   tables, 346

Mean deviation measures, 236-237
Mean measurement, 228-230
Measurement, defined, 274
Measures, area, 306-317
   capacity, 330-339
   linear, 274-299
   mass, 235-339
   statistical, 228-237
   tables, 343-346
   volume, 319-328
   weight, 335-339
Median measurement, 231-233
Metric-English equivalents, area measure, 314
   capacity, 333
   linear, 296
   tables, 342-346
   volume, 326
   weight, 337
Metric feeler gauge, 303
Metric units, advantages of using, 293
   area measure, 310-314
   capacity measure, 331
   frequently used, table, 292
   linear measure, 291-299
   tables, 343-346
   volume measure, 322-328
   weight measure, 335
Millilitres, 330, 331
Minuend, 14
Mixed numbers, adding fractions and whole numbers, 72, 96
   dividing fractions, 106-110
   expressing as fractions, 58
   expressing fractions as, 59
   multiplying fractions, 93, 96
   subtracting fractions, 80-83, 96
Mode measurement, 233-235
Multiplicand, 21
Multiplication, common fractions, 90-102, 105
   compound numbers, 283
   decimal fractions, 152-157
   fractions, 56
   whole numbers, 21-32
Multiplication table, 22
Multiplier, 21

N
Nonterminating decimal, 134
Numbers, denominate, 274
   whole, 1-6
Numerator, 34
   fractions, 52

O
Operations, order of, 43-50

P
Parentheses, use, 178-179
Percentage, defined, 212
Percents, 207-218
   as common fractions, 211
   as decimal fractions, 211
   common fractions as, 210
   decimal fractions as, 209
   definitions, 208
   finding percentages, 213-216
   types of problem, 212-213, 222-224
Perfect squares, 183
Place value, 2
Plus sign, 8
Power, meaning, 176, 178
Powers and roots, 176-187
Powers of ten, 126
   dividing by, 168
   multiplying by, 156
Prime factor, 66
Prime number, 66
Product, 21

Q
Quotient, 34
   trial, 36

R
Radical symbol, 180
   expressions enclosed within, 182
Range measures, 235-236
Rate, percentage, 212
   finding percent, 215
   finding percentage by, 213
Reading bar graphs, 241
Reading decimal fractions, 127-129
Reading line graphs, 254-261
Reading whole numbers, 4-6
Regrouping process, 15
Remainder, in subtraction, 14
Roots, 176
   combining operations, 186
   computing, 183
   description, 180
Rounding decimal fractions, 132

S
Scales, graphs, 240
Short multiplication, 22
Spread of data measures, 228

Square root, 176
    combining operations, 186
    computing, 183
    description, 180
Squares, percent defined by, 208
Statistical measures, 228-237
    mean, 228-230
    mean deviation, 236-237
    median, 231-233
    mode, 233-235
    range, 235-236
Statistics, defined, 228
Straight-line graphs, 254
    drawing, 264-265
Subtraction, common fractions, 79-89, 108
    compound numbers, 281-282
    decimal fractions, 143-145, 157, 170
    whole numbers, 14-20
Subtrahend, 14
Sum, 8
Surface measure, 306-317
    English units, 306-310
    metric-English equivalents, 314-317
    metric units, 310-314
Symbols, division, 34
    fractions, 51
    metric units, 291-292
    parentheses, 178
    percent, 208

Symbols, division (con't)
    radical, 180
    surface measure, 310
    use in formulas, 43

T

Tables, area measure, 343
    capacity measure, 345
    linear measure, 342
    multiplication, 22
    volume measure, 343
    weight (mass) measure, 346
Temperature, formula, 46
Ten, powers of, 126
    dividing by, 168
    multiplying by, 156
Terminating decimal, 134
Terms of fraction, 52
Times sign, 21
Torque, 121
Trial quotients, 36
Troy weights, 335

V

Vertical scale, graphs, 240
Volume, finding, 176
Volume measure, English units, 319-322
    metric units, 322-324
    tables, 343

W

Weight, English and metric units of, 335-339
    tables, 346
Welded support bar, 175
Whole numbers, 1
    addition, 8-13, 72
    division, 33-42
    expanding, 3
    multiplication, 21-32
    order of operations, 43-50
    place value, 2
    reading and writing, 4-6
    subtraction, 14-20, 80
Work, formula, 272
Writing decimal fractions, 129
Writing metric units, 293-294
Writing whole numbers, 4-6

X

X-axis scale, 240

Y

Y-axis scale, 240

Z

Zero, as dividend, 34
    as divisor, 34
    in multiplier, 25